Physics of
CRYSTALLINE
DIELECTRICS
Volume 2
Electrical Properties

Physics of
CRYSTALLINE DIELECTRICS
Volume 2
Electrical Properties

I. S. Zheludev
Institute of Crystallography of the
Academy of Sciences of the USSR
Moscow, USSR

Translated from Russian by
Albin Tybulewicz
Editor, *Soviet Physics — Semiconductors*

ℚ PLENUM PRESS • NEW YORK-LONDON • 1971

Prof. Ivan Stepanovich Zheludev was born in 1921. He was graduated in 1950 from the Physics Department (Radiophysics Section) of the Moscow State University. The following year he joined the Crystallography Institute of the Academy of Sciences of the USSR and in 1960 became Head of the Laboratory of Electrical Properties of Crystals at this Institute. He was awarded the degree of Candidate of Physico-Mathematical Sciences in 1954 and won his doctorate in 1961. His scientific research has always been concerned with the properties of dielectrics (particularly ferroelectrics, piezoelectrics, and pyroelectrics) in relation to their structure and symmetry. Professor Zheludev has published extensively. He is the coauthor (with V. M. Fridkin) of the monograph *Photoelectrets and the Electrophotographic Process* published in 1960 and translated into English (Consultants Bureau, New York, 1961). Zheludev taught first as a lecturer and then as a professor at the Moscow State University (1955–61); at the Indian Institute of Sciences, Bangalore (1962–63); and at the Voronezh University (1964–66). Since 1966 he has held the temporary appointment of Deputy General Director of the International Atomic Energy Agency in Vienna.

The original Russian text, published in one volume by Nauka Press in Moscow in 1968, has been corrected by the author for this edition. Volume 1 of the present edition is comprised of pages 1–236, inclusive, of the original edition and Volume 2 of pages 237–425, inclusive. Pages 426–461 are not included in the present edition. The English translation is published under an agreement with Mezhdunarodnaya Kniga, the Soviet book export agency.

Желудев Иван Степанович

ФИЗИКА КРИСТАЛЛИЧЕСКИХ ДИЭЛЕКТРИКОВ

FIZIKA KRISTALLICHESKIKH DIELEKTRIKOV

Library of Congress Catalog Card Number 79-138522
ISBN-13: 978-1-4615-8986-0   e-ISBN-13: 978-1-4615-8984-6
DOI: 10.1007/ 978-1-4615-8984-6

© 1971 Plenum Press, New York
Softcover reprint of the hardcover 1st edition 1971
A Division of Plenum Publishing Corporation
227 West 17th Street, New York, N. Y. 10011

United Kingdom edition published by Plenum Press, London
A Division of Plenum Publishing Company, Ltd.
Donington House, 30 Norfolk Street, London W.C. 2, England

Preface to the American Edition

Research in solid-state physics in general and in the physics of dielectrics in particular has grown rapidly in scope and quantity in the last twenty-five years. In the fifties and early sixties, there was an upsurge of interest in ferroelectricity, piezoelectricity, and pyroelectricity. The classical physics of dielectrics, represented by books of H. Fröhlich, C. P. Smyth, G. I. Skanavi, and A. von Hippel, is now unthinkable without ferroelectricity. The structure and properties of ferroelectrics have been described in a number of books and reviews, including those of W. Känzig, H. D. Megaw, F. Jona and G. Shirane, W. J. Merz and E. Fatuzzo.

The present work deals with the physics of crystalline dielectrics and is based on the investigations carried out by scientists throughout the world. But, understandably, the emphasis is on the research done in the USSR, particularly in the author's laboratory. A special feature of this two-volume treatise is the prominent place given to the symmetry and structure of dielectrics and to the importance of spontaneous electric polarization in many properties of crystals. In fact, these aspects take up the whole of the first volume. The second volume is concerned mainly with various properties and phenomena whose nature is illustrated by considering specific crystals. Thus, for example, the phenomena of polarization, piezoelectricity, electrostriction, etc., are first discussed in detail. Then follow descriptions of these phenomena in specific compounds. This approach seems to be more systematic and it is hoped that the book will be useful to a wide range of readers.

I. S. Zheludev

Preface to the Russian Edition

This book is an extended treatment of various aspects of the structure and electrical properties of dielectric crystals. It is the outgrowth of several lecture courses given by the author in the last decade to research workers, postgraduate students, and undergraduates at the Crystallography Institute of the Academy of Sciences of the USSR, at the Moscow and Voronezh State Universities, and at the Indian Institute of Science at Bangalore.

Since the main stress is laid on the properties of crystals, the traditional subjects of the physics of isotropic dielectrics (polarization, electrical conductivity, and dielectric losses) are supplemented by chapters or sections dealing with pyroelectric and piezoelectric properties, electrocaloric phenomena, electrostriction, etc. Since the spontaneous polarization plays an exceptionally important role in many properties of crystals, the atomic and domain structure of dielectrics and the theory of spontaneous polarization are given special treatment in this book.

Sections 1 and 2 of Chapter III were written mainly by A. S. Sonin; Section 1 of Chapter VI was contributed by V. V. Gladkii. G. A. Zheludeva and L. P. Pashkova helped the author considerably in the preparation of the manuscript for press.

The author is most grateful to all those who helped him in writing and revising the manuscript as well as in publishing this book.

I. S. Zheludev

General Introduction

The physics of crystalline dielectrics deals with all the phenomena and properties which are related in some way to the electric polarization of crystals. The physics of dielectrics covered in this book includes the phenomena and properties resulting from the application of an external electric field or those resulting from the spontaneous polarization observed in some crystals, as well as the electric polarization phenomena due to mechanical strain or stress (piezoelectric polarization).

Even until fairly recently, the physics of dielectrics consisted mainly of the following subjects: electric polarization, electrical conductivity, dielectric losses, and electric strength of isotropic dielectrics (capacitor and insulation materials) widely used in technology.

Recent developments in technology have been characterized by extensive use of crystals. Thus, the successful development of quantum electronics, piezoelectric techniques, electroacoustics, metrology, etc., would be impossible without the use of crystals. Moreover, crystals are widely used in the well established branches of electrical and radio engineering, such as capacitor manufacture and electric insulation technology. These applications of crystals have led to the emergence and rapid development of new subjects in the physics of dielectrics: pyroelectricity, piezoelectricity, electrostriction, etc. But only when we are armed with full knowledge of the properties of a given dielectric, including its polarizability, electrical conductivity, electric strength, pyroelectric and piezoelectric properties, and elastic and electrostriction parameters, can we realize all the possible technological applications of this material.

The establishment of the physics of dielectrics as an independent branch of physics is closely related to the work of Soviet scientists, particularly A. F. Ioffe, I. V. Kurchatov, Ya. I. Frenkel' (J. Frenkel), B. M. Vul, A. K. Val'ter, P. P. Kobeko, G. I. Skanavi, N. P. Bogoroditskii, and A. A. Vorob'ev. The development of the physics of dielectrics during the last three or four decades has resulted in the establishment of industrial and research organizations and institutions which are supplying Soviet electrical and radio engineering industries with the necessary dielectric materials.

The modern physics of dielectrics is basically the physics of ferroelectrics and antiferroelectrics, as well as of piezoelectric and pyroelectric materials. The milestones in the development of this new physics of dielectrics are the investigations of ferroelectrics, carried out in the nineteen-thirties by I. V. Kurchatov and his colleagues, the studies of B. M. Vul and others which led to the discovery (in 1944) of the most important ferroelectric material, barium titanate, as well as the work of V. L. Ginzburg on the theory of ferroelectricity.

Important results relating to piezoelectric and ferroelectric crystals were obtained by A. V. Shubnikov and his colleagues. G. A. Smolenskii and his team had many successes in the preparation of new oxygen-octahedral ferroelectrics and in the study of their properties. A key place in the investigations of the structure of ferroelectrics is occupied by the work of G. S. Zhdanov and his colleagues.

The physics of dielectrics is part of solid-state physics. Consequently, further development of the physics of dielectrics will depend on the extensive use of diffraction methods for structure investigation (x-ray, neutron, and electron diffraction methods), resonance methods for the investigation of the structure of crystals and the fields in them (nuclear magnetic resonance, quadrupole resonance, Mössbauer effect), and on the many varieties of spectral investigations covering a wide range of electromagnetic wavelengths from the visible right down to the millimeter region.

The promising directions of investigation in the physics of dielectrics are those concerned with materials exhibiting simultaneously several properties: these materials include photoconducting ferroelectric and piezoelectric substances, semiconducting piezoelectric materials, materials exhibiting magnetic and electric ordering of their structure, and so on.

Introduction to Volume 2

The information contained in Volume 1 on the crystallography and structure of crystals, as well as on their spontaneous polarization and its nature, will be used in this second volume to provide a comprehensive and physically clear description of the electrical properties of dielectrics.

In Volume 2, we shall consider electric polarization (Chap. VII), electrical conductivity and dielectric losses (Chap. VIII), piezoelectric effect and electrostriction (Chap. IX).

In addition to the general problems, which are usually discussed in the physics of isotropic dielectrics, there are many particular features of the polarization, electrical conductivity, and dielectric losses of ferroelectrics, which are associated with their spontaneous polarization and domain structure. The phenomena of pulse switching (polarization reversal) in ferroelectrics, their nonlinear properties, etc. are specific to this class of materials.

Piezoelectric properties are encountered mainly in dielectric crystals. These properties accompany other properties of dielectric crystals more frequently — and are more important — than is usually assumed. The phenomenon of electrostriction also deseves special attention because electrostrictive strains in ferroelectric crystals can be of practical importance.

At the rear of the volume, between pages 614 and 615, the reader will find a foldout containing a conversion table for the Shubnikov, Schoenflies, and International crystallographic notation systems.

Contents of Volume 2

Contents of Volume 1

Chapter VII

Electric Polarization

Introduction

Electric polarization is the most important property of dielectrics. It is now assumed that dielectrics (including ferroelectric crystals) can exhibit the following types of polarization:

1) polarization associated with the displacement of electrons relative to the nuclei of their atoms (electronic displacement polarization or, simply, electronic polarization);

2) polarization associated with the displacement of ions in a crystal lattice relative to one another (ionic displacement polarization or, simply, ionic polarization);

3) polarization associated with the orientation of existing permanent dipole moments (thermal orientational polarization or, simply, orientational polarization);

4) polarization associated with the motion of weakly bound ions (thermal ionic polarization).

The first two types of polarization, involving the elastic displacement of charged particles, are usually both called the displacement polarization. The third and fourth types of polarization originate in the thermal motion of particles in a dielectric, and their establishment is a relatively slow process. They are known jointly as the relaxational polarization.

In addition to these four types of polarization, heterogeneous dielectrics may exhibit interfacial polarization and, in strong electric fields, high-voltage polarization can also be observed.

The presence of a domain structure in ferroelectrics is responsible for several special features of their electric polarization. Of particular interest are the dependences of the polarization on the field frequency and intensity, the pulse polarization processes, and the correlation of the polarization processes with changes in the domain structure of ferroelectrics. Of cardinal importance are the dielectric nonlinearity of ferroelectrics and dielectric hysteresis. The polarization associated with the state of the surface of a crystal and the presence of surface layers are also special properties of ferroelectrics.

In this chapter, we shall consider some ferroelectrics and antiferroelectrics, and discuss in detail the aspects of their electric polarization mentioned in this introduction.

§1. Electric Fields in Dielectrics

A. Polarization in Anisotropic Media. Crystals of dielectrics become polarized when placed in an external electric field. The electric field E in the interior of a dielectric can be determined only if its polarization P is taken into account. The state of a dielectric is represented, in general, by the vectors E and P as well as the vector D which is variously known as the electric induction, electric flux density, or electric displacement. The electric induction D is due to free charges and the polarization P is due to bound charges (if this definition is used the normal component of the polarization vector P at the boundary of a dielectric with vacuum is numerically equal to the density of bound charges). The difference between free and bound charges* determines the electric field intensity E. The vectors D, P, and E are related by the following expressions

$$\left.\begin{array}{l} \boldsymbol{P} = \alpha \boldsymbol{E}, \\ \boldsymbol{D} = \boldsymbol{E} + 4\pi \boldsymbol{P}, \\ \boldsymbol{D} = \varepsilon \boldsymbol{E}, \\ (\varepsilon = 1 + 4\pi\alpha), \end{array}\right\} \qquad \text{(VII.1)}$$

where α is the polarizability of the dielectric and ε is its permittivity.

---

* Bound charges belong to the particles of which a given dielectric is composed. These charges cannot be removed from the dielectric without destroying it. Here, the term free charges does not apply to particles in a dielectric.

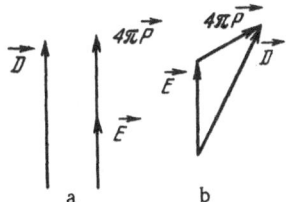

Fig. 152. Vectors **D**, **E**, and **P** in isotropic (a) and anisotropic (b) dielectrics.

The quantities ε and α are scalars for an isotropic medium. In the case of crystals (and all other anisotropic bodies), the quantities α and ε, which relate the polar vectors, are polar tensors of the second rank $[\alpha_{ij}]$, $[\varepsilon_{ij}]$. The relationship between the vectors **D**, **E**, and **P** is shown graphically in Fig. 152. We note that the polarization of a dielectric produces surface charges which are of sign opposite to the sign of charges associated with the electric induction vector. Bearing this in mind, as well as the relationship in Eq. (VII.1), the direction of the vector **P** within a dielectric can be selected so as to point from negative to positive charges. (Lines of force of the free-charge field are always assumed to start at positive charges and are directed toward negative charges.)

It follows from Onsager's symmetry principle that, in the absence of magnetic fields (including spontaneous fields), the tensors $[\alpha_{ij}]$ and $[\varepsilon_{ij}]$ in static electric fields are symmetrical.

We shall consider first the process of polarization described by a symmetrical polar tensor $[\alpha_{ij}]$. The principal coefficients α_i and ε_i are always positive, i.e., the polarizability of dielectrics is always positive. This allows us to represent the process of polarization in dielectrics by characteristic ellipsoidal surfaces. For crystals of the cubic system the dielectric ellipsoid is a sphere. In this case, the vectors **D** and **E** always coincide in direction and the permittivity is a scalar. In the case of crystals of moderately high degrees of crystal symmetry, the characteristic surface is an ellipsoid of revolution, whereas for crystals of low-symmetry systems the surface is a general ellipsoid.

Let us consider the equation of the characteristic surface

$$\varepsilon_1 x^2 + \varepsilon_2 y^2 + \varepsilon_3 z^2 = 1 \qquad\qquad (VII.2)$$

for the tensor $[\varepsilon_{ij}]$ and the equation for the characteristic surface

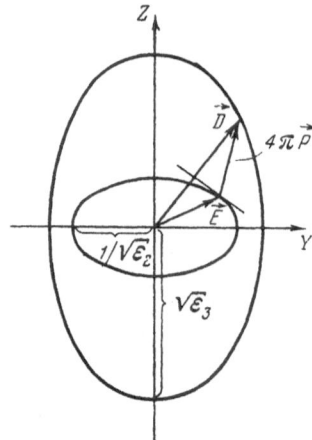

Fig. 153. Plotting of the vectors D and $4\pi P$ from a known value of the vector E in an anisotropic dielectric. The illustration shows the special case of a section of an ellipsoid of revolution (whose axis coincides with the Z axis) by the YZ plane.

of its reciprocal $[\varepsilon_{ij}]^{-1}$:

$$\frac{x^2}{\varepsilon_1} + \frac{y^2}{\varepsilon_2} + \frac{z^2}{\varepsilon_3} = 1. \qquad (VII.3)$$

Using these equations, we can easily determine [6] the magnitudes and directions of D and P for any given direction of E (Fig. 153). The same dielectric ellipsoidal surfaces can be used to find E and P when D is given. It is evident from Fig. 153 that all three vectors D, E, and P coincide only along the principal axes of the ellipsoid. Along these axes the polarization is represented by the principal polarizabilities α_1, α_2, α_3, or by the principal permittivities ε_1, ε_2, ε_3. Similar characteristic surfaces can be constructed using the tensors $[\alpha_{ij}]$ and $[\alpha_{ij}]^{-1}$.

If we take an arbitrary direction we find that, in general, the vectors D, E, and P do not coincide in direction and the meaning of the quantities ε and α is no longer as simple as for isotropic media and crystals. By analogy with the optics of anisotropic media, we shall introduce the concept of permittivity ε_E along the direction of E and permittivity ε_D along the vector D. The value of ε_E indicates how many times the vector E is shorter than the projection of the vector D along the direction of E. The value of ε_D shows how many times the vector D is longer than the projection of E along the direction of D.

Using Eqs. (VII.2) and (VII.3) ($X \to E_1$, $Y \to E_2$, $Z \to E_3$) and specifying the directions of the vectors E and D in terms of the direction cosines c_i of the angles made by these vectors with the axes of the coordinate systems E_1, E_2, E_3, and D_1, D_2, D_3, respectively, we obtain formulas which can be used to calculate ε_E and ε_D from known principal permittivities ε_1, ε_2, and ε_3:

$$\left.\begin{aligned}\varepsilon_E &= c_1^2 \varepsilon_1 + c_2^2 \varepsilon_2 + c_3^2 \varepsilon_D, \\ \frac{1}{\varepsilon_D} &= \frac{c_1^2}{\varepsilon_1} + \frac{c_2^2}{\varepsilon_2} + \frac{c_3^2}{\varepsilon_3}.\end{aligned}\right\} \qquad \text{(VII.4)}$$

These two expressions can be used to plot, for every crystal, double surfaces which represent a permittivity along a given direction.

In order to find which of the quantities, ε_E or ε_D, is yielded by the experimental data, we must consider in more detail the orientations of the vectors D, E, and P in a parallel-plate capacitor (uniform-field case).

Although the tangential component of the electric field, E_t, is continuous, it does not follow that the field E in a dielectric is equal to the electric induction $D = 4\pi\sigma$. Only some of the lines of force of D pass through the dielectric and determine the value of E, whereas other lines of D end at the bound charges σ' (Fig. 154b).

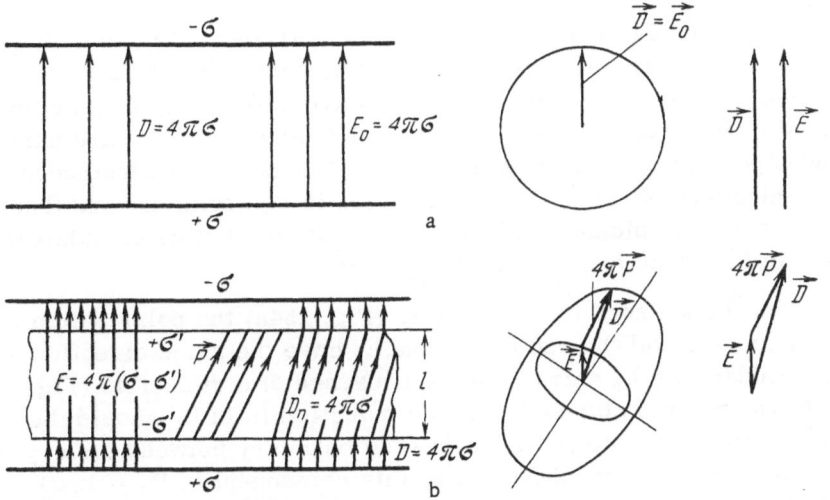

Fig. 154. Directions and magnitudes of the vectors D, E, and 4πP in a parallel-plate capacitor containing vacuum (a) and a crystalline plate (b).

Thus, $E = 4\pi(\sigma - \sigma')$, where $\sigma' = P_n$ is the normal component of P. The results of graphical determinations of D and P from a given value of E, using dielectric ellipsoids, are also given in Fig. 154.

We shall now find the permittivity ε_E along the field E. By definition, $D_n = \varepsilon_E E$ and hence

$$\varepsilon_E = \frac{D_n}{E} = \frac{E_0}{E}. \tag{VII.5}$$

The permittivity ε of a dielectric in a capacitor can be determined experimentally from the ratio of the capacitances with (C) and without (C_0) the dielectric:

$$\frac{C}{C_0} = \frac{\varepsilon}{1}$$

(the permittivity ε of vacuum is 1). Next, using the relationship $\sigma = CV$, $V = El$ (l is the thickness of the dielectric in the capacitor), we obtain

$$E_0 / E = \varepsilon. \tag{VII.6}$$

Comparing Eqs. (VII.5) and (VII.6), we find that $\varepsilon_E \equiv \varepsilon$, which shows that the experimentally determined value of ε is the permittivity along the direction of the vector E, i.e., it is the quantity ε_E.

Figure 154 shows the capacitor electrodes with and without a dielectric plate between them. It is assumed that the capacitor electrodes and the crystal plate faces are sufficiently large compared with the thickness of the crystal l and that there are narrow gaps between the crystal plate and the capacitor electrodes (the thickness of these gaps is assumed to be negligibly small compared with the thickness of the crystal plate). Let us postulate that each of the electrodes has a charge $\pm \sigma$.

In the absence of a dielectric (Fig. 154a) the polarization is equal to zero and the vectors D and E_0 have the same direction and magnitude: $D = E_0 = 4\pi\sigma$. In the presence of a dielectric (Fig. 154b) the normal component of the induction field D_n should be unchanged in the gap of the uncharged boundary between the crystal plate and vacuum (the value of this component is $D_n = 4\pi\sigma$), but the vector D is not directed perpendicularly to the electrodes. On the other hand, the tangential component E_t should also be con-

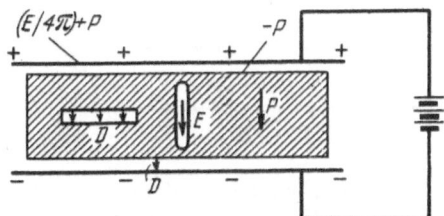

Fig. 155. Definitions of the vectors **D** and **E**.

tinuous at this boundary. Since in vacuum $E_t = 0$, the vector E in
the dielectric should be perpendicular to the electrodes. The same
result can be deduced from the postulate that the field in such a
capacitor should be uniform and hence the equipotential surfaces
should be parallel and the field should be perpendicular to the
electrodes.

B. Electrostatic Field Equations in Any
Medium and Some Applications of These Equations.
We have considered so far the principal macroscopic relationships
describing the processes of polarization of crystals and taken into
account their anisotropy. We shall now discuss briefly some of
the most typical and important examples of the polarization of
solids.

The total system of equations for the electrostatic field in
any system is of the form

$$E = -\operatorname{grad} \varphi, \qquad D = \varepsilon E,$$
$$\operatorname{div} D = 4\pi\rho, \qquad D_{2n} - D_{1n} = 4\pi\sigma, \qquad \text{(VII.7)}$$

where E is the electric field intensity, D is the electric induction, and
ρ and σ are, respectively, the volume and surface densities of free
charges. Analysis of the system (VII.7) shows easily that the electric field inside a long needle-shaped cavity, perpendicular to the
electrodes of a parallel-plate capacitor, is equal to the average
macroscopic field E. On the other hand, the field inside a disk-
shaped cavity, parallel to the capacitor plate, is equal to the electric flux density (or induction) D (Fig. 155). The second conclusion follows directly from the continuity of the normal components
of the vector D on an uncharged surface [fourth expression in the
system (VII.7)]. The fourth and second expressions in (VII.7) yield
the continuity of the tangential components of the vector E at an

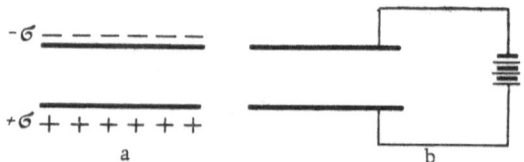

Fig. 156. Capacitor in the case of constant flux density (charge) conditio..s (a) and under constant field conditions (b).

uncharged boundary. This explains the first conclusion. However, we must point out that this behavior of the normal and tangential components of E and D shows that the lines of force of these vectors are refracted at the boundary.

In some cases, we have to find the fields in a parallel-plate capacitor. In such problems we must distinguish two basically different states of the capacitor. One of these states is characterized by a constancy of the electric induction (D = const) and the other is characterized by a constancy of the electric field (E = const). The former case represents a charged open-circuited capacitor (Fig. 156a), and the latter case represents a capacitor connected to a voltage source (Fig. 156b). When a dielectric of permittivity ε is introduced into the first capacitor the field intensity in this dielectric is ε times lower than the electric induction (which is equal to the field intensity in vacuum E_0), i.e., $E = E_0/\varepsilon$. The introduction of a dielectric into the second capacitor has two effects: first, it polarizes the dielectric, which can be represented by a surface density of bound charges σ'; secondly, the initial charge density σ increases by an amount σ' in order to maintain the constant potential difference between the plates (we recall that this capacitor is connected to a constant-voltage source). This is equivalent to a new value of the electric induction D', corresponding to the free charge density $\sigma + \sigma'$ and the original field intensity in the dielectric ($E = E_0$).

In the case of a two-layer capacitor, filled with dielectrics of permittivities ε_1 and ε_2, which are parallel to the capacitor plates, we find that if D = const the electric fields in layers 1 and 2 are given by the expressions

$$E_1 = \frac{E_0}{\varepsilon_1}, \qquad E_2 = \frac{E_0}{\varepsilon_2}.$$

In a capacitor with D = const and containing two dielectrics (permittivities ε_1 and ε_2), separated by a boundary perpendicular to the plates, the field intensity in both dielectrics is the same: $E_1 = E_2$. This follows from the postulate that all parts of the capacitor electrode must be at the same potential, i.e., $E_1 d = E_2 d$.

Since

$$E_1 = \frac{4\pi\sigma_1}{\varepsilon_1}, \qquad E_2 = \frac{4\pi\sigma_2}{\varepsilon_2},$$

where σ_1 and σ_2 are the densities of charges above dielectrics 1 and 2 in this capacitor, we find that

$$\frac{\sigma_1}{\sigma_2} = \frac{\varepsilon_1}{\varepsilon_2}. \qquad\qquad \text{(VII.8)}$$

We can use a similar procedure to find the fields in layered dielectrics in the case when E = const.

Let us now consider a case of practical importance, which is the field in a uniformly polarized sphere. In such a sphere the centers of gravity of positive and negative charges (which coincide before polarization) are separated by a distance $2l$, as shown in Fig. 157. The external field of a polarized sphere is identical with the field of a dipole consisting of two charges $\pm q$ (charges of positive and negative spheres, respectively), separated by a distance $2l$, i.e., the field of a dipole whose moment is $p = 2ql = 2VeNl$ (V is the volume of the sphere, e is the charge of a single particle, and N is the number of particles per unit volume). Using the electrostatic relationship for the determination of the potential of a dipole,

$$\varphi = \frac{p \cdot R}{R^3},$$

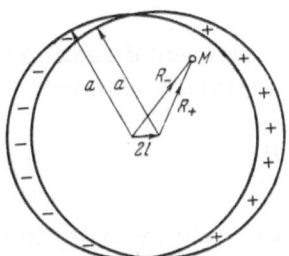

Fig. 157. Calculation of the field in a uniformly polarized sphere.

we obtain

$$\varphi_{ext} = 2eNV \frac{l \cdot R}{R^3} \qquad (R > a), \qquad (VII.9)$$

where R is the radius vector drawn from the center of the sphere to the point under consideration M, and a is the radius of the sphere.

On the other hand, the electric moment per unit volume of the sphere (i.e., the polarization) $p = \Sigma e_i R_i$ is zero before the application of an electric field and $p = \Sigma e_i (R_i \pm l) = 2eNl$ after such application. Using this equation, we obtain from Eq. (VII.9):

$$\varphi_{ext} = V \frac{P \cdot R}{R^3} \qquad (R > a). \qquad (VII.10)$$

The potential at internal points in the sphere (R ≤ a) is given by a similar formula but, in this case, V does not represent the volume of the whole sphere but only of that part which is closer to the center than the point M under consideration, i.e., in this case, we must substitute into Eq. (VII.10)

$$V = \frac{4\pi}{3} R^3,$$

so that we obtain

$$\varphi_{int} = \frac{4\pi}{3} R^3 \frac{P \cdot R}{R^3} = \frac{4\pi}{3} P \cdot R \qquad (VII.11)$$

It is evident that the expressions (VII.10) and (VII.11) become identical at R = a, i.e., the potential φ of a polarized sphere is a continuous function of the coordinate.

Finally, the field intensity within a polarized sphere is found to be

$$E_{int} = - \nabla \varphi_{int} = - \frac{4\pi}{3} \nabla (P \cdot R)(R \leqslant a).$$

Since the vector P is constant in magnitude and direction it follows that $\nabla(P \cdot R) = P$ and, therefore, we finally obtain

$$E_{int} = - \frac{4\tau}{3} P \, (R \leqslant a). \qquad (VII.12)$$

It follows from this relationship that the field intensity in a uniformly polarized sphere is constant in magnitude and direction at all its internal points.

C. Average Macroscopic (E) and Internal or Effective (F) Fields. When we consider the polarization of atoms and molecules (microprocesses), we find it useful to employ the concepts of the internal or effective electric field. This is because, in the macroscopic approach, the field intensity in a dielectric is understood to be the average field E which ignores the atomic structure of the dielectric. However, the process of polarization of molecules and atoms are not governed by this average field but by the internal (effective) field F. We can easily see that the fields E and F differ, if only because the average field E is calculated taking into account the field at all points (not only, for example, the field at the centers of atoms) for all charges (including charges at a given point), whereas the effective field F applies only to specific points and when it is calculated the field of charges at that point is ignored and only the field of charges outside this point is taken into account.

Let us consider a polarized dielectric. In order to determine the field intensity F at some point O in this dielectric, we shall describe around this point a sphere S such that its radius is very small compared with the macroscopic inhomogeneities of the dielectric and very large compared with the interatomic distances (Fig. 158). The field F at the point O can be calculated as the sum of: 1) the field E_1 of all the charges outside the sphere S; 2) the field E_2 of all the charges within the sphere S, with the exception of the charges at the point O. Thus,

$$F = E_1 + E_2. \tag{VII.13}$$

In order to calculate the field E_1 we shall imagine that matter is removed from the sphere S. The field at the point O is then equal

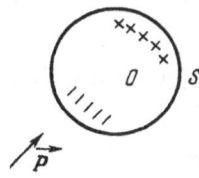

Fig. 158. Calculation of the microscopic (effective) field in a dielectric.

to E_1. Charges which give rise to this field are outside the sphere S and, therefore (neglecting the atomic structure), we can replace all the molecules in the dielectric by a continuous medium with an electric moment **P**. Bearing in mind the conditions governing the selection of the size of the sphere S, we find that the polarization **P** of the dielectric surrounding the sphere is constant in magnitude and direction. Thus, E_1 is equal to the field intensity in a spherical cavity cut within a uniformly polarized dielectric. Before the removal of matter from this spherical cavity, the field in this uniformly polarized dielectric (including the field within the sphere S) was uniform and equal to the average macroscopic field E. The removal of the charges within the sphere S obviously reduced the value of E by an amount equal to the field of a uniformly polarized sphere, which is given — in accordance with Eq. (VII.12) — by $-4\pi P/3$. Thus,

$$E_1 = E - \left(-\tfrac{4\pi}{3}\,P \right) = E + \tfrac{4\pi}{3}\,P. \qquad (VII.14)$$

The second term in this expression is usually called the Lorentz field.

The value of the field E_2, i.e., of the field of charges lying within the sphere S, can be calculated only it the structure of a given dielectric is taken into account. However, we can show that, in some cases, $E_2 = 0$. This applies particularly to dielectrics with a completely disordered distribution of atoms or molecules (gases, nonpolar liquids).

In fact, if all the positions of any specific molecule within the sphere S are equiprobable and independent of the positions of other molecules, the average value of the field of any molecule, measured at the center of the sphere O, is zero. However, the assumption of complete absence of correlation between the positions of different molecules is equivalent to the assumption that the dimensions of molecules are negligibly small and that the dipolar interaction between the molecules can be ignored. It is found that the field E_2 is equal to zero for cubic monatomic and diatomic crystals. However, in general, it is fairly difficult to calculate the field E_2.

If we restrict ourselves to gases, nonpolar liquids, and cubic monatomic and diatomic crystals we find that

$$F = E + \tfrac{4\pi}{3}\,P.$$

The knowledge of the effective field **F** allows us to relate the microscopic characteristics (the polarizabilities of ions, atoms, or molecules η) to the macroscopic polarizability of a dielectric α. On the one hand,

$$P = N\eta F = \alpha_0 F = \alpha_0 E + \frac{4\pi}{3}\alpha_0 P, \qquad (VII.15)$$

where N is the number of structure units per unit volume. On the other hand, according to Eq. (VII.1),

$$P = \alpha E. \qquad (VII.16)$$

where α is the macroscopic polarizability. Comparing Eqs. (VII.15) and (VII.16), we find that:

$$\alpha = \frac{\alpha_0}{1 - \frac{4\pi}{3}\alpha_0}. \qquad (VII.17)$$

Using the relationship between the polarizability α and permittivity ε

$$\varepsilon = 1 + 4\pi\alpha = 1 + \frac{4\pi\alpha_0}{1 - \frac{4\pi}{3}\alpha_0}, \qquad (VII.18)$$

we find that simple transformations yield the formula

$$\frac{\varepsilon - 1}{\varepsilon + 2} = \frac{4\pi}{3}\alpha_0 = P_0, \qquad (VII.19)$$

where P_0 is the specific polarization.

The last expression, written in the form

$$\frac{M}{\rho}\frac{\varepsilon - 1}{\varepsilon + 2} = \frac{4\pi}{3}N_0\eta = P, \qquad (VII.20)$$

where M is the molecular weight, ρ is the density of the dielectric, N_0 is Avogadro's number, and η is the polarizability of a molecule, is known as the Clausius—Mosotti formula, in which P is the molecular polarization.

Since the permittivity ε and the refractive index n are – in many cases – related by $\varepsilon = n^2$, Eq. (VII.20) can be rewritten in the form

$$\frac{M}{\rho} \frac{n^2 - 1}{n^2 + 2} = \frac{4\pi}{3} N_0 \eta = P_E. \qquad \text{(VII.21)}$$

The quantity P_E in the above expression is known as the molar refraction, and Eq. (VII.21) is known as the Lorentz–Lorenz formula.

The Lorentz field $4\pi P/3$ in Eq. (VII.14) takes into account the mutual interaction of dipoles which are induced in a polarized dielectric. The smallness of this interaction is typical of weakly polarizable media. For such media the fields $\mathbf{E}$ and $\mathbf{F}$ differ very little from one another [$\alpha_0 \approx \alpha$, as shown by Eq. (VII.17)]. We must point out also that many assumptions have been made in the derivation of the Clausius –Mosotti formula and, therefore, its validity is very limited. In particular, the field E_2 in liquid and solid dielectrics having polar structures is known to be finite and this indicates that the formulas just given are inapplicable to polar liquids and solid dielectrics.

§2. Polarization Associated with the Elastic Displacements of Charges

A. Electronic Displacement Polarization. This type of polarization is common to all dielectrics. It is due to an elastic displacement of the least strongly bound electrons (mainly valence electrons) in an atom or ion. After such a displacement, the centers of gravity of the electron cloud and the nucleus of an atom (ion) no longer coincide, i.e., the atom (ion) acquires a dipole moment. The time necessary for the establishment of the electronic displacement polarization is comparable with the period of electromagnetic oscillations in the optical band and amounts to 10^{-14}-10^{-15} sec.

Every type of polarization can be represented by its own polarizability. Generally, the polarizability of atoms increases with increasing number of electrons and the lengths of the radii of the electron orbits. The polarizabilities of elements belonging to the same group in the periodic table increase with the atomic number, i.e., as the total number of electrons and the radius of the outermost orbit become larger. Elements in the same period have polarizabilities which may increase or decrease with increasing atomic number, depending on which of the two effects predominates:

the increase in the number of electrons or the decrease in the orbit radius.

In some cases, the electronic polarizability of ions is of importance. This polarizability is approximately of the same nature as the electronic polarizability of corresponding atoms. The most stable are those ions whose electron shells have the configuration of the inert-gas shells. It is interesting to note that the electronic polarizability of such ions decreases with increasing atomic number. A high polarizability is exhibited by oxygen ions O^{2-}, whose electron-shell configuration is very far removed from the shell configuration of inert gases.

The electronic polarizability of ions may be compared with the ionization potential. A high ionization potential is a consequence of strong binding between an electron and the nucleus of its atom. Therefore, the higher the ionization potential, the smaller should be the polarizability.

The electric moment per unit volume, which defines the polarization, depends not only on the induced moment in each elementary particle but also on the number of such particles per unit volume. Therefore, the polarizability can be represented by the ratio of the polarizability of a particle η and the cube of the ionic radius r^3. Dielectrics with the highest values of this ratio should have high polarizabilities and small ions; the permittivity of such dielectrics should be high. High values of the ratio η/r^3 are typical of Ti^{4+}, Pb^{2+}, and O^{2-} ions.

The permittivity of a dielectric ε may be due to various types of polarization. However, in the optical range of frequencies it is almost entirely due to the electronic polarizability. In this case, the specific refraction is given by the relationship [see Eq. (VII.19)]

$$\frac{n^2 - 1}{n^2 + 2} = \frac{4\pi}{3} N_i \eta_i, \qquad\qquad \text{(VII.22)}$$

where N_i is the number of atoms of an i-th type per unit volume and η_i is the electronic polarizability of an atom of the i-th type.

Using the experimentally determined values of the refractive indices and Eq. (VII.22), we can calculate quite accurately the values of the electronic polarizability of many crystals. The polarizabilities found in this way for various atoms and ions, de-

TABLE 37. Electronic Polarizability of Ions (10^{-24} cm^3) Based on Pauling's Data (P) [63] and on the Results of Tessman, Kahn, and Shockley (S) [72]

			He	Li^+	Be^{2+}	B^{3+}	C^{4+}
P			0.201	0.029	0.008	0.003	0.0013
S				0.03			
	O^{2-}	F^-	Ne	Na^+	Mg^{2+}	Al^{3+}	Si^{4+}
P	3.88	1.04	0.390	0.179	0.094	0.052	0.0165
S	2.4	0.652		0.41			
	S^{2-}	Cl^-	Ar	K^+	Ca^{2+}	Sc^{3+}	Ti^{4+}
P	10.2	3.66	1.62	0.83	0.47	0.286	0.185
S	5.5	2.97		1.33	1.1		0.19
	Se^{2-}	Br^-	Kr	Rb^+	Sr^{2+}	Y^{3+}	Zr^{4+}
P	10.5	4.77	2.46	1.40	0.86	0.55	0.37
S	7.0	4.17		1.98	1.6		
	Te^{2-}	I^-	Xe	Cs^+	Ba^{2+}	La^{3+}	Ce^{4+}
P	14.0	7.10	3.99	2.42	1.55	1.04	0.73
S	9.0	6.44		3.34	2.5		

termined using the D lines of sodium, are listed in Table 37. The discrepancies between some values listed in the table (obtained by different authors) are due to differences in the environment of ions (i.e., crystals) used in the relevant experiments. It is evident from Table 37 that the electronic polarizabilities of all atoms and ions are of the order of 10^{-24} cm^3.

The electronic polarizability can be calculated easily using the classical theory. If the forces binding an electron to an atom are such that the vibrations of the electron are harmonic, classical mechanics predicts that the frequency of these vibrations is

$$\omega_0 = \sqrt{\frac{\beta}{m}},$$

where β is the elastic force constant and m is the mass of the electron. Resonance absorption takes place at this frequency. The average displacement of the electron, $\bar{x}$, caused by an electric field, F, can be found from the condition

$$eF = \beta x = m\omega_0^2 x.$$

The static electronic polarizability is

$$\eta_e^{st} = \frac{p}{F} = \frac{e\bar{x}}{F} = \frac{e^2}{m\omega_0^2}. \tag{VII.23}$$

We can show that, in the simplest case (for example, in the case of a hydrogen atom), the elastic force constant is given by

$$\beta = m\omega_0^2 = \frac{e^2}{r^3}, \tag{VII.24}$$

where r is the radius of the electron orbit. Comparing Eqs. (VII.23) and (VII.24), we obtain

$$\eta_e^{st} = r^3, \tag{VII.25}$$

which shows that the value of η_e^{st} is of the order of 10^{-24} cm^3.

The electronic polarizability depends on the frequency. In order to find the frequency dependence, we shall consider the equation of motion of an electron in an electric field:

$$m\ddot{x} + \beta x = eF. \tag{VII.26}$$

If the applied field is alternating at a frequency $\omega/2\pi$ we have

$$F = F_0 e^{i\omega t}.$$

Solving Eq. (VII.26) for x and multiplying the solution by e, we obtain a relationship between the induced electric moment p and the field F in the form

$$xe = p = \eta_e F,$$

in which the electronic polarizability η_e is given by the relationship

$$\eta_e = \frac{e^2}{m\,(\omega_0^2 - \omega^2)}. \tag{VII.27}$$

In the optical range of frequencies the dispersion of η_e of the majority of dielectrics is slight.

A quantum-mechanical treatment of the electronic polarizability gives the following expression [2]:

TABLE 38. Permittivity ε and Refractive Index n
of Some Solid Dielectrics Exhibiting Electronic
Polarization

Dielectric	ε	n	n^2
Sulfur	3.8—4.1	1.92	3.85
Naphthalene	2.50	1.58	2.50
Paraffin	2.0—2.3	—	—
Polystyrene	2.5—2.9	1.55	2.4

$$\eta_e = \frac{e^2}{m} \sum \frac{f_{ij}}{\omega_{ij}^2 - \omega^2} , \qquad (\text{VII.28})$$

where f_{ij} is the oscillator strength given by the relationship

$$f_{ij} = \frac{2\omega_{ij} m (x_{ij})^2}{\hbar} , \qquad (\text{VII.29})$$

and $2\pi\hbar$ is the Planck constant.

An order-of-magnitude estimate of the electronic polarizability of simple atoms, obtained using Eq. (VII.28), gives a value
of η_e close to that found using the classical theory [Eq. (VII.27)].

The electronic displacement polarization can be observed
in its pure form (i.e., in the absence of other types of polarization) in those crystals whose lattices consist of nonpolar molecules (diamond, naphthalene, paraffin) and in some solid organic
polymers (polystyrene, polyethylene, and others). The process of
polarization of such dielectrics in static and low-frequency fields,
as well as in fields of optical frequencies, is described satisfactorily by the Clausius–Mosotti formula [Eqs. (VII.19), (VII.20)].
These dielectrics satisfy at all frequencies the relationship $n^2 =$
ε [compare with the Lorentz–Lorenz equation (VII.21)], which
can be seen quite clearly from Table 38.

Dielectrics which exhibit only the electronic polarization
usually have a weak temperature dependence of the permittivity.
The only exceptions to this rule are regions near the temperatures of phase transitions, depolymerization, decomposition, etc.
In the absence of such phenomena the permittivity ε of a dielectric varies only because of variation of the number of particles

per unit volume, i.e., because of the expansion or compression of a crystal due to temperature variations.

The temperature dependence of the permittivity is given by the quantity

$$\beta_\varepsilon \equiv \frac{1}{\varepsilon} \frac{d\varepsilon}{dT}. \tag{VII.30}$$

Applying the Clausius—Mosotti formula to such dielectrics and denoting the volume expansion coefficient by Δ_V, we obtain

$$\beta_\varepsilon = \frac{1}{\varepsilon} \frac{d\varepsilon}{dT} = -\frac{(\varepsilon-1)(\varepsilon+2)}{3\varepsilon} \Delta_V. \tag{VII.31}$$

Assuming, in the first approximation, that $\Delta_V = 3\alpha_L$, where α_L is the linear expansion coefficient, and taking $\varepsilon \approx 2\text{-}2.5$, we obtain

$$\beta_\varepsilon \cong -2\alpha_L.$$

For many dielectrics $\alpha_L \approx 10^{-4} \text{ deg}^{-1}$. This means that the temperature coefficient β_ε for dielectrics which exhibit only the electronic polarizability is also of the order of $-10^{-4} \text{ deg}^{-1}$. It is interesting to note that this temperature coefficient is very small (when the temperature of a dielectric is raised by 100°C, ε changes only by 0.1) and is negative: the expansion of a dielectric during heating reduces the number of structure units per unit volume, i.e., it reduces its density.

B. Ionic Displacement Polarization. The ionic displacement polarization (also known simply as the ionic polarization) is due to the displacement of ions relative to one another and is exhibited most clearly by crystals with ionic or predominantly ionic (heteropolar) nature of binding. As already mentioned, such crystals exhibit ionic and electronic displacement polarizations. However, the electronic displacement polarization cannot alone explain the experimentally observed values of the permittivity of some ionic crystals. Thus, the refractive index of sodium chloride is $n = 1.5$, and, therefore, $n^2 = 2.25$, but the static permittivity is $\varepsilon_s = 5.62$. This difference between the values of the static ($\varepsilon_s = 5.62$) and optical ($\varepsilon_\infty = 2.25$) permittivities must be due to the ionic displacement polarization which can be represented by the polarizability of an ion pair η_i.

An analysis of the polarization due to elastic displacements of ions, carried out in the same way as in the case of deformation of electron orbits, gives the following relationship for ionic crystals of the NaCl type:

$$\eta_i = \frac{e^2}{\beta'},\tag{VII.32}$$

where β' is the elastic force constant of ions. We shall express the constant β' in terms of the frequency of free vibrations of the lattice:

$$\omega_0 = \sqrt{\frac{\beta'}{m'}}.$$

In this relationship m' denotes the reduced mass of two ions, whose individual masses are m_1 and m_2; the reduced mass is given from the relationship

$$\frac{1}{m'} = \frac{1}{m_1} + \frac{1}{m_2}.$$

It thus follows from Eq. (VII.32) that

$$\eta_i = \frac{e^2}{\omega_0^2} \left(\frac{1}{m_1} + \frac{1}{m_2} \right).\tag{VII.33}$$

The ionic polarizability per unit volume, α_i, is found assuming that the volume per one "molecule" of sodium chloride is $2a^3$ (a is the distance between the nearest neighbors):

$$\alpha_i = \frac{e^2}{2\omega_0^2 a^3} \left(\frac{1}{m_1} + \frac{1}{m_2} \right).\tag{VII.34}$$

Measurements on ionic crystals carried out at optical frequencies yield, in practice, only the electronic polarizability because natural frequencies of ionic displacements lie in the infrared region. The permittivity of an ionic crystal at radio frequencies and in static fields ε_s is thus due to the electronic (α_e) and ionic (α_i) polarizabilities:

$$\varepsilon_s = 1 + 4\pi(\alpha_e + \alpha_i).$$

Representing the contribution of the electronic polarization ($1 + 4\pi\alpha_e$) by n^2 (the permittivity at optical frequencies), we obtain finally

$$\varepsilon_s = n^2 + \frac{2\pi e^2}{\omega_0^2 a^3} \left(\frac{1}{m_1} + \frac{1}{m_2} \right).\tag{VII.35}$$

This expression was obtained first by Born and is known as the Born formula. An estimate of the contribution of the ionic displacement polarization to the permittivity of NaCl, obtained using the Born formula, gives

$$\varepsilon_i = 2.7,$$

which is close to the difference $\varepsilon_i = 5.62 - 2.25 \approx 3.4.$

Born made many assumptions in the derivation of his formula. In particular, he ignored the difference between the values of the effective field at low (quasistatic) and optical frequencies and did not take into account the difference between the effective and average macroscopic fields. These assumptions are equivalent to the postulate that the ions are rigid (undeformable) and non-overlapping. An improved variant of the Born formula was derived by Szigeti [71a, 71b]:

$$\varepsilon_s - n^2 = \varepsilon_i = \left(\frac{n^2 + 2}{3} \right) \frac{2\pi (e^*)^2}{\omega_t^2 a^3} \left(\frac{1}{m_1} + \frac{1}{m_2} \right), \qquad (VII.36)$$

where ω_t is the natural frequency (eigenfrequency) of the transverse vibrations in the optical mode and e^* is the effective charge of ions.

A comparison of the values of ε, calculated using the Born formula, and of the effective charges, calculated using the Szigeti formula, is given in Table 39 for alkali halide crystals. It is evident from this table that the Born formula gives more or less

TABLE 39. Experimental Check of the Formulas of Born and Szigeti for Some Alkali-Halide Crystals

Crystal	$n^2 = \varepsilon_\infty$	Experimental value ε_{exp}	Value of ε calculated using Born's formula	Ratio e^*/e calculated using Szigetis's formula
LiF	1.92	9.27	8.1	0.87
NaCl	2.25	5.62	5.3	0.74
KCl	2.13	4.68	4.3	0.80
NaBr	2.62	5.99	9.5	0.69
NaI	2.91	6.60	8.0	0.71
KBr	2.33	4.78	10.0	0.76
KI	2.69	4.94	10,5	0.69
RbCl	2.19	5.0	—	0.84
RbBr	2.33	5.0	10.0	0.82
RbI	2.63	5.0	11.0	0.89

satisfactory results for some of these crystals. For other crystals the discrepancy between the calculated and experimental values is larger. The table also gives the ratio of the effective and purely ionic charges, obtained from the Szigeti formula, which shows that the bonds in these crystals have some degree of covalence (homopolarity). A comparison of the calculated values of the lattice energies of these crystals (carried out on the assumption that they are purely ionic) with the experimental energies should yield values of the ratio $e*/e$ close to unity. The discrepancy between these expected values and those listed in Table 39 has not yet been accounted for.

Calculations of the temperature dependences of the permittivity of ionic crystals are somewhat more difficult than the corresponding calculations for substances exhibiting only the electronic polarization. The electronic polarization of ionic crystals increases with increasing temperature because of decreasing density of the crystal. The ionic polarization should (for the same reason) decrease with increasing temperature; however, it should also increase because of the decreasing value of the elastic force constant. Experiments show that the permittivity ε of ionic cubic crystals increases almost linearly with increasing temperature and that this increase can be represented by the coefficient $(1/\varepsilon) \cdot d\varepsilon/dT$. Assuming, in the first approximation, that the elastic-force constant varies with temperature only due to a change in the distance between the ions and using a power law for the repulsive energy of the electron shells, we can find the derivative of the ionic polarizability with respect to the temperature, i.e., we can find $d\varepsilon/dT$. This simple calculation gives satisfactory results even when we ignore changes in the ionic polarizability due to changes in the amplitude of ion vibrations with increasing temperature. A calculation of this type, reported by Skanavi [17], shows that the coefficient $(1/\varepsilon) \cdot d\varepsilon/dT$ for NaCl should be $+36 \times 10^{-5}$ deg^{-1}, which is in good agreement with the experimental value of $+34 \times 10^{-5}$ deg^{-1}. The corresponding values for KCl are $+24 \times 10^{-5}$ deg^{-1} and $+30.3 \times 10^{-5}$ deg^{-1}. The coefficient $(1/\varepsilon) \cdot d\varepsilon/dT$ of many crystals is positive and (as in the case of dielectrics with a purely electronic polarizability) of the order of 10^{-4} deg^{-1}. Some dielectric crystals (for example, TiO_2) have negative values of the temperature coefficient.

Concluding our discussion of the ionic displacement polarization, we must draw attention to the high permittivity of some oxides with the rutile structure (TiO_2, SnO_2, PbO_2). Crystals of these oxides have high refractive indices ($\varepsilon = n^2 = 4.4$-7.3). The permittivity of one of the modifications of TiO_2 (rutile) along a fourfold axis of symmetry is 171 and the corresponding permittivities of SnO_2 and PbO_2 are 24 and 26, respectively. The ionic nature of the lattices of these crystals, combined with the relatively weak temperature dependence of the permittivity ε, shows that the principal types of polarization in these crystals are the electronic and ionic displacement polarizations. This can be said also about some of the nonferroelectric perovskites.

The hypothesis that the large difference between the radio-frequency values of n^2 and ε of rutile-type crystals can be due to the "open" crystal structure of rutile (Fig. 19b) has not been confirmed since the high eigenfrequencies of the lattice vibrations indicate that the ionic displacement polarization in these crystals cannot be appreciably higher than, for example, in alkali halides. In view of this, other attempts have been made to explain the nature of the high polarizability of these crystals. The most important results have been obtained by Skanavi [17, 18]. He has calculated in detail the effective field exerted on the lattice ions in rutile structures. He has used the method of structure coefficients (Chap. VI, §3) and has taken into account the influence of a large number of neighboring ions (over 150). Skanavi regards a crystal lattice as an array of point charges and assumes that an external electric field produces an array of point dipoles. Even this relatively rough approximation explains successfully the observed discrepancies. The main results of Skanavi's calculations can be formulated as follows.

1. In the case of a purely electronic polarization in the rutile lattice, the Lorentz field must be supplemented by an internal (effective) field E_2 [Eq. (VII.13)] which is due to the polarization of surrounding ions of the same sign (the "like" ions). In the rutile lattice this field is directed opposite to the external field.

2. The ionic displacement polarization, which appears at infrared and lower frequencies, gives rise to an additional (effective) field due to the surrounding ions of the opposite sign (the "unlike"

ions). In the rutile lattice this field coincides in direction with the
external field. Therefore, even a small ion displacement is sufficient to increase the permittivity well above the value of the square
of the refractive index. It is obvious that this increase becomes
larger for larger values of the electronic polarizability (larger
values of the refractive index). Thus, in spite of the fact that the
power-frequency (50 or 60 cps) permittivity of rutile and similar
crystals is very much higher than the square of the refractive index, the nature of the polarizability of such crystals is predominantly electronic, i.e., the ionic polarizability is small.

In conclusion, we must point out that the temperature dependence of the permittivity of rutile has a minimum near 0°C.
The origin of this minimum is as follows: at low temperatures
the reduction in the density of the crystal with increasing temperature predominates over the increase in the ionic polarizability, but
at high temperatures the reverse is the case.

The elastic displacement polarization can also be in the form
of rotation (through small angles) of strongly bound dipoles in polar dielectrics subjected to an external field. However, this type
of polarization can be considered more conveniently in a general
discussion of the polarization of polar dielectrics (see the following section).

§3. Relaxational Polarization

Relaxational polarization is primarily the polarization due to
changes in the orientation of free or weakly bound dipoles in polar dielectrics and the polarization due to the motion of weakly
bound ions. These two types of polarization are closely related
to the thermal motion of particles (they depend on this motion) and,
therefore, they are called thermal orientational and thermal ionic
polarizations. In addition to the thermal polarization, we shall
consider briefly two other types of polarization of the relaxational
type: interfacial and high-voltage polarization.

A. Thermal Orientational Polarization. Some
dielectrics contain molecules or other structure units which have
a permanent dipole moment μ. Molecules which have such a moment are called polar. If polar molecules are free, an external
electric field tends to orient them along its direction and the de-

gree to which this orientation is achieved is governed by thermal
motion (such motion disorients these free polar molecules).

Free polar molecules can be found in gases and liquids. Debye [5] was the first to investigate the polarization of dielectrics
containing polar molecules. In particular, he explained the high
permittivity of water ($\varepsilon = 81$) and other liquids by considering the
orientational polarization, which is important in static fields and
at radio frequencies right up to the centimeter range. The results of investigations of the polarizability of free polar molecules are useful also in the description of solid dielectrics and,
therefore, we shall consider them briefly.

Let us consider the effect of thermal motion on free molecules. The potential energy of a molecule with a permanent dipole moment μ in a field E is of the form

$$U = -\mu E = -\mu E \cos \theta, \qquad \text{(VII.37)}$$

where θ is the angle between the dipole moment and the direction
of the field. If N is the number of molecules per unit volume and
$\overline{\cos \theta}$ is the average value of this factor for a system in thermal
equilibrium, we find that the polarization of a dielectric is given
by $P = N\mu \overline{\cos \theta}$.

According to the Boltzmann distribution law, the relative
probability that the dipole moment of a given molecule lies within a solid angle element $d\Omega$ is proportional to exp $(-U/kT)$ and,
therefore,

$$\overline{\cos \theta} = \frac{\int e^{-U/kT} \cos \theta \, d\Omega}{\int e^{-U/kT} \, d\Omega} \qquad \text{(VII.38)}$$

(k is the Boltzmann constant and T is the absolute temperature).
Expressed in polar coordinates, the above expression becomes

$$\overline{\cos \theta} = \frac{\int\limits_0^\pi 2\tau \cdot \sin \theta \cdot \cos \theta \cdot e^{\mu E \cos \theta/kT} \cdot d\theta}{\int\limits_0^\pi 2\pi \cdot \sin \theta \cdot e^{\mu E \cos \theta/kT} \cdot d\theta}.$$

Using the notation cos θ = x and $\mu E/kT = a$, we find that

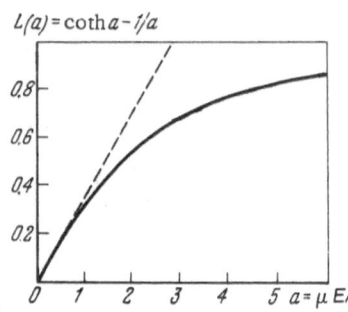

Fig. 159. Langevin function L(a).

$$\overline{\cos\theta} = \frac{\int\limits_{-1}^{+1} e^{ax}x\,dx}{\int\limits_{-1}^{+1} e^{ax}\,dx} = \frac{d}{da}\ln\int\limits_{-1}^{+1}e^{ax}dx = \coth a - \frac{1}{a} \equiv L(a), \quad \text{(VII.39)}$$

where L(a) is the Langevin function.

The Langevin function is shown graphically in Fig. 159. It follows from this figure that, in the case of free molecules ($\mu E \ll kT$, i.e., when $a \ll 1$), we have

$$L(a) \cong \frac{a}{3} = \frac{\mu E}{3kT}. \qquad \text{(VII.40)}$$

We can easily see that, for the usual values of the moments of polar molecules ($\mu \sim 10^{-18}$ cgs esu) in moderate fields (E = 3000 V/cm) at room temperature (kT = 4×10^{-14} erg), we have $a = \mu E/kT \approx 1/4000$, i.e., the condition that a is small compared with unity is satisfied very well. In such cases, the polarizability of a dielectric is

$$P = N\mu\overline{\cos\theta} = \frac{N\mu^2 E}{3kT}, \qquad \text{(VII.41)}$$

and hence the value of the orientational polarizability of a dielectric is

$$\alpha_{or} = \frac{N\mu^2}{3kT}. \qquad \text{(VII.42)}$$

The room-temperature orientational polarizability of a single polar molecule

$$\eta_{or} = \frac{\mu^2}{3kT} \qquad \text{(VII.43)}$$

is of the order of 10^{-23} cm^{-3} (for the values of μ just given), which is close to the values of the electronic and ionic polarizabilities.

Using η_0 to denote the sum of the electronic and ionic polarizabilities, $\eta_e + \eta_i$, and including the orientational polarizability $\eta_{or} = \mu^2/3kT$, we find that the total polarizability of a structure unit is

$$\eta = \eta_0 + \frac{\mu^2}{3kT}. \qquad \text{(VII.44)}$$

Using the relationships

$$\varepsilon = 1 + 4\pi\alpha, \quad \alpha = N\eta \quad (E = F),$$

we obtain the expression

$$\varepsilon - 1 = 4\pi N \left(\eta_0 + \frac{\mu^2}{3kT}\right), \qquad \text{(VII.45)}$$

which is known as the Langevin-Debye formula.

In this formula, all types of polarization are regarded as additive and the interaction between the dipoles is ignored. In spite of these deficiencies, the Langevin-Debye formula is of considerable importance in the interpretation of molecular structure. Using the experimental dependence of η, or of the molar polarizability P [Eq. (VII.20)], on the reciprocal of temperature, we can use Eqs. (VII.44) or (VII.35) to find graphically the value of the dipole moment μ. The dipole moments μ of HCl, HBr, HI, and H_2O, determined in this way, are, respectively, 1.03, 0.79, 0.38, and 1.87×10^{-18} cgs esu.

The dipole moments of polar molecules are frequently given in Debye units, equal to 10^{-18} cgs esu. The order of magnitude of a dipole moment, expressed in Debye units, is given by the product of the electronic charge and the interatomic distance.

So far we have considered the processes of polarization of polar molecules under conditions such that the energy of a dipole is much less than the energy of the thermal motion ($\mu E \ll kT$). This is the case which is encountered most frequently in practice.

However, in strong fields and at low temperatures we can reach the other limiting case $\mu E \gg kT$. In this case, $L(a)$ is close to unity (Fig. 159) and the expression for the polarization becomes

$$P = N\mu.$$

This means that the directions of the dipole moments of all molecules are almost coincident with the field direction, i.e., we are close to saturation. However, such states are encountered relatively infrequently.

It is natural to expect that molecules can rotate relatively easily in gases and liquids. In such cases, the average potential energy of a single dipole is the same for all directions in spite of some tendency of the dipoles of neighboring molecules to orient themselves along certain directions relative to one another. The situation is quite different in a molecular crystal, where the average potential energy of a polar molecule depends, in general, on the direction of the dipole with respect to the symmetry elements of the crystal structure. The internal crystal field, which is due to the interaction between molecules, is usually temperature-dependent. A crystal (depending on its symmetry) may have one or several directions along which the dipole energy has relative minima. Such directions are called the equilibrium (stable) directions. The potential barriers between the equilibrium directions of polar molecules are, in general, very high and they hinder free rotation of these molecules, even at temperatures close to the melting point. However, nonpolar molecules of nearly spherical shape can rotate relatively freely. Thus, nonpolar solid methane (CH_4), which consists of highly symmetrical molecules, can exhibit a relatively free molecular rotation. Solid hydrogen behaves similarly.

If the energy of thermal motion in a dielectric containing polar molecules is sufficient to overcome the potential barriers between the equilibrium directions, the molecules can "jump" from one equilibrium direction to another. The external field distorts the potential barriers and establishes a preferential direction for the random thermal jumps between the fixed equilibrium directions. The field thus polarizes a dielectric.

At certain critical temperatures the properties of polar dielectrics may change rapidly. To explain these changes, Pauling suggests that above a critical temperature molecules can rotate freely. However, the current view is that such free rotation does not occur in the majority of solids. The changes occurring in a dielectric at a critical temperature should be regarded as a transition from an ordered to a disordered distribution of dipoles.

As mentioned in Chap. I, crystalline dielectrics which do not exhibit the multiple-cell structure can exist in three states. The first of these states is polar, i.e., it is the state in which the dipoles of all the molecules are oriented in parallel. In this state a dielectric exhibits spontaneous polarization. The second state is called polar-neutral: in this state the orientations of the dipoles are such that there is no net spontaneous polarization. Dielectrics which do not contain polar molecules (or other polar units) have the third — nonpolar — type of structure. In the case of dielectrics exhibiting multiple-cell structure we can also have states with "linear" and "volume" antiparallel orientations of the dipoles. These two states are achieved easily by pairing dipoles in such a way that their orientations are antiparallel. The first of these states corresponds to pairing along a unique direction in a crystal and the second corresponds to pairing along several equivalent directions in polar-neutral structures.

Each of these states has its own dipole energy. Let us assume that the state of lowest energy, existing at a temperature close to absolute zero, is polar and fully ordered. This state then corresponds to the minimum value of the interaction energy. The next higher energy state may be the linear antipolar state. A transition to this state implies the appearance of multiple-cell structure: a unit cell in a dielectric now contains two dipoles, in contrast to the presence of only one dipole per unit cell in the polar state. If the antipolar state is reached at some finite temperature, it means that some disorder appears in a crystal at this temperature. At the critical temperature corresponding to a transition to this disordered state the difference between the energies of the dipoles oriented along the two antiparallel directions is equal to zero since both these directions are equiprobable. In this new state we can have the short-range order, characterizing the orderly distribution of neighbors, but the long-range order is only partial since the antipolarization direction is unique. The long-range order disappears completely only at the melting point.

This type of partial disordering in a solid dielectric is not the only possible one. Transitions can also take place between polar states in which the direction of polarization changes (i.e., the multiple-cell structure does not appear); we can also have transitions in which the dipole moments are destroyed, etc. We must bear in mind that transitions in solids, which give rise to or alter the dipole moments, should be matched by corresponding changes in the symmetry of a crystal, discussed in detail in §1 of Chap. IV.

The ordering of dipoles in a solid dielectric at low temperatures may mean that the application of an electric field can produce only an elastic displacement (rotation through a small angle) of these dipoles. This type of polarization is typical of solid dielectrics with "frozen" dipoles. An analysis shows (see, for example, Skanavi [17]) that the polarizability due to an elastic displacement (rotation) of the polar molecules η_μ is given by

$$\eta_\mu = \frac{\mu_0^2 \sin^2\theta}{|U_0|} , \qquad (VII.46)$$

where θ is the angle between the direction of the electric field and quasielastic restoring force; U_0 is the binding energy of a polar molecule due to these restoring forces. The binding energy U_0 for "frozen" dipoles is usually fairly high and, therefore, the values of η_μ are usually low. The time for the establishment of the polarization of elastically coupled molecules is governed by the period of their natural vibrations and is usually of the order of 10^{-13} sec.

At higher temperatures, the application of a field to a solid polar dielectric may alter the orientations of the dipoles; this may result in a rise of the permittivity ε_s with increasing temperature. It can be shown (see, for example, Fröhlich [23]) that, in the case of a transition from a polar to antipolar state, the permittivity due to the reorientation of dipoles is given by the relationship

$$\varepsilon_s - \varepsilon_\infty = \frac{3\varepsilon_s}{2\varepsilon_s + n^2} \cdot \frac{4\pi N\mu\mu^*}{3kT} \, 4w\,(1 - w), \qquad (VII.47)$$

where 2μ is the dipole moment of a unit cell; $2\mu^*$ is the moment of a spherical region surrounding this cell; w is the probability that half the dipoles form the antipolar structure; N is the number of dipoles per unit volume. At low temperatures $w \to 0$, and the contribution of the dipole reorientation becomes negligibly small. At a critical temperature T_0 we have $w = 1/2$ (fully anti-

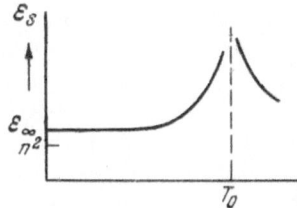

Fig. 160. Temperature dependence
of the permittivity ε_s of a polar
dielectric solid near the point of
transition from the ordered to the
disordered state (schematic representation).

parallel structure) and Eq. (VII.47) becomes

$$\varepsilon_s - \varepsilon_\infty = \frac{3\varepsilon_s}{2\varepsilon_s + n^2} \cdot \frac{4\pi N \mu \mu^*}{3kT}, \qquad \text{(VII.48)}$$

which is similar to Kirkwood's formula for the polarization of polar liquids. We note that in Eqs. (VII.47) and (VII.48) the symbol
ε_∞ does not represent the permittivity at optical frequencies (n^2)
but the permittivity due to the electronic displacements and the
elastic dipole displacements (rotation of the dipoles through small
angles).

Thus at $T = 0$ a "frozen" dielectric consisting of polar molecules (with any type of ordering) has a permittivity ε_∞ which is
somewhat higher than n^2. As the temperature increases, the static
permittivity ε_s increases, at first slowly, and then the rate of rise
becomes greater on approach to T_0 (provided the critical temperature T_0 lies below the melting point). At temperatures $T > T_0$ the
value of ε_s decreases with increasing temperature and shows no
appreciable changes at the melting point (Fig. 160).

We shall now consider some specific cases in order to com-
pare the experimental data with the general ideas on the polarization of solid crystalline dielectrics containing dipoles. We shall
deal with four different cases of polarization in crystals containing
polar molecules or polar radicals.

In the first case, ε decreases rapidly on transition from the
liquid to solid state. Such behavior is exhibited, for example, by
nitrobenzene, nitromethane, and aniline. These substances consist
of polar molecules, and they exhibit clearly the thermal orientational
polarization in the liquid state. The strong fall of ε at the freezing
point indicates strong binding of polar molecules in the crystal lattice,
which prevents reorientation of the dipoles, i.e., which shows that
the rotational degrees of freedom, responsible for the thermal

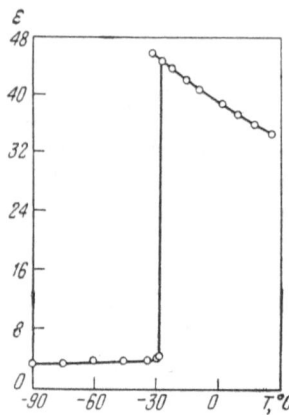

Fig. 161. Temperature dependence
of the permittivity of nitromethane.

orientational polarization, are "frozen out." The temperature dependence of the permittivity for this type of material (nitrobenzene) is shown in Fig. 161.

In the second case, the permittivity ε of a solid crystal is
similar to ε in the liquid state. A typical example of a material
of this kind is ice (the permittivity ε of ice is 80–81, i.e., it is practically equal to ε of water). The nature of the polarization of ice

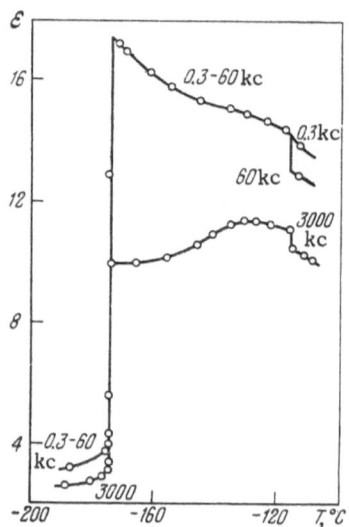

Fig. 162. Temperature dependence
of the permittivity of hydrochloric
acid.

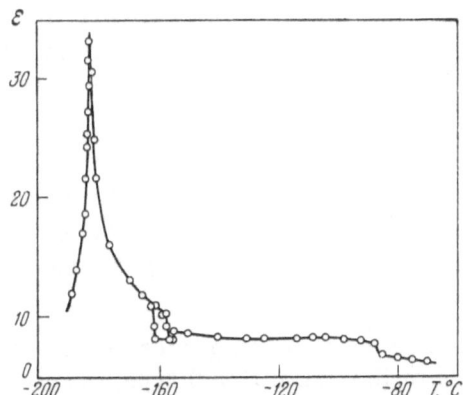

Fig. 163. Temperature dependence of the permittivity of HBr.

indicates that the polar radicals (OH hydroxyl groups) can rotate
in the ice lattice. We note that the density of ice is less than the
density of water, which indicates that water molecules are more
tightly packed than the ice molecules.

In the third case, the transition of a dielectric from the liquid to the solid state does not alter appreciably the permittivity ε.
However, at some temperature, which is much lower than the melting point, the permittivity ε falls rapidly. Such a temperature dependence of ε is exhibited by solid hydrochloric acid (Fig. 162),
whose freezing point is −114°C and which exhibits a sudden drop
of ε at −174°C. The latter temperature corresponds to a transition from the cubic to the orthorhombic modification and is accompanied by a decrease in the specific heat of crystalline hydrochloric acid. In such cases we speak of the "orientational melting" of a crystal at a temperature corresponding to a rapid fall of
ε (for HCl this temperature is −174°C).

In the fourth case, the ordering produces a gradual fall of ε.
This fall continues until the dipole rotation is completely "frozen."
An example of this type (Fig. 163) is HBr (its crystallization temperature is −160°C and the ordering temperature is −185°C).
Similar behavior below the ordering temperature is exhibited also by hydrogen sulfide, dimethyl sulfate $[(CH_2O)_2SO_4]$, and by other organic compounds (methyl alcohol, Halowax, etc.).

The orientational polarization is observed also in crystals containing water of crystallization. However, the permittivity ε of such crystals is usually lower than that of polar crystals.

It is evident from Figs. 160–163 that in all four cases the permittivity of the liquid state decreases with increasing temperature. This is in satisfactory agreement with the relationships governing the polarization of polar liquids [see Eq. (VII.43)], which indicates that the thermal orientational polarizability is inversely proportional to the temperature.

We have pointed out that the processes of polarization of polar dielectrics are of the relaxational type. The establishment or destruction of the polarization in such dielectrics is associated with the thermal motion of dipoles (or particles carrying a dipole moment), is of the probability type, and requires a certain time known as the relaxation time τ. We can easily see that the value of τ is given by the relationship

$$\tau = \frac{e^{U/kT}}{2\nu},\qquad\qquad\text{(VII.49)}$$

where ν is the frequency of natural vibrations of a polar particle; U is the height of the potential barrier separating its two equilibrium positions, k is the Boltzmann constant; T is the absolute temperature. At a frequency ν each dipole attempts to overcome the potential barrier twice in every period ($\tau' = 1/2\nu'$ where τ' is the time between "attempts"). The product of the time between "attempts" and exp (U/kT) gives Eq. (VII.49); the exponential factor is the reciprocal of the probability of a transition in accordance with the Boltzmann statistics (the higher the probability of a transition, the shorter is the time taken in transition). Similar considerations apply also to the thermal ionic polarization. The concept of the relaxation time will be discussed in greater detail in Chap. VIII.

So far, we have considered mainly the temperature dependence of the polarization of molecular crystals formed from polar liquids. The nature of the polarization and ordering in such crystals can be understood on the basis of the polarization of polar liquids. The nature of the polarization (in the external electric field) of other crystalline polar dielectrics, namely, ferroelectrics and antiferroelectrics, is completely different; these crystals exhibit many special properties because of the presence of domains (see §§4-9 in the present chapter).

B. Thermal Ionic Polarization. The thermal motion of particles causes not only the polarization of weakly bound or free dipoles but also the polarization of weakly bound ions. A solid dielectric frequently contains weakly bound ions which may be detached from their regular sites by the thermal motion, and they can migrate through a crystal. Such migration depends on the structure of the dielectric.

An external field establishes a preferential direction in the random thermal motion of such ions and thus causes more ions to move along this direction. In a solid dielectric, the motion of weakly bound ions is greatly restricted. Therefore, the motion of ions along the direction of the field produces an asymmetry in the volume-charge distribution and, after some time, this asymmetry becomes compensated by the back diffusion of ions. Thus, under steady-state conditions an electric moment (polarization) and a counterfield are established in each element of volume of the dielectric. This form of polarization is known as the thermal ionic polarization.

In calculations of the value of the thermal ionic polarizability it is usual to assume that, in the absence of an external field, a

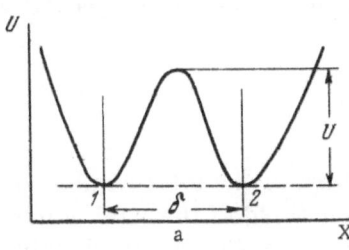

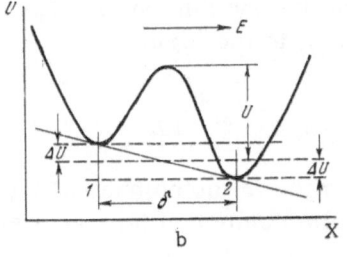

Fig. 164. Potential energy minima for two positions of an ion in the absence (a) and presence (b) of an external electric field.

weakly bound ion can occupy two equiprobable positions 1 and 2
with identical potential energies (Fig. 164a). The probability of a
transition of an ion from position 1 to position 2 (or conversely)
depends on the height of the potential barrier U and on the temperature, i.e., according to the Boltzmann statistics, the probability is governed by the quantity exp (−U/kT), where k is the
Boltzmann constant, and T is the absolute temperature. If an ion
is fixed in position 1 or 2 it then vibrates at frequency ν, which is
governed by the elasticity of the binding forces.

In the absence of external forces, transitions along all directions in space are equally probable and one–sixth of the total
number n_0 of weakly bound ions undergoes transitions along the
positive directions of three equivalent axes (it is assumed that the
structure is isotropic and the motion is random). The number of
attempts to overcome the potential barrier along a given direction is $\nu n_0/6$ and the number of ions n which do overcome the barrier is given by the expression

$$n = \frac{n_0}{6}\,\nu e^{-U/kT}. \tag{VII.50}$$

The application of an external field E alters appreciably the
situation. We shall assume that this field is directed along the X
axis. Such a field deforms the total potential energy as shown in
Fig. 164b. A change in the potential energy by an amount ΔU over
a distance $\delta/2$ (for a charge e) is

$$\Delta U = \frac{eE\delta}{2}.$$

Let us assume that n_1 is one–third of the total number of
ions occupying position 1 and n_2 is one–third of the total number of
ions occupying position 2. Then $n_1 + n_2 = n_0/3$. A vanishingly
small change dn_1 in a very short time interval dt can be represented,
in the first approximation (for U > kT), in the form

$$dn_1 = \left(-n_1\nu e^{-\frac{U-\Delta U}{kT}} + n_2\nu e^{-\frac{U+\Delta U}{kT}}\right)dt. \tag{VII.51}$$

Let us now assume that a reduction in the number of ions
occupying position 1 (an increase in the number of ions at posi-
tion 2) is Δn. It is obvious that $\Delta n = n_0/6 - n_1 = n_2 - n_0/6$, so that

$n_2 - n_1 = 2\Delta n$. Substituting Δn in Eq. (VII.51) and considering initially only weak fields which satisfy $\Delta U \ll kT$ and the approximate equality $\exp(\pm U/kT) \cong 1 \pm \Delta U/kT$, we obtain

$$\frac{d(\Delta n)}{dt} = -2\Delta n v e^{-\frac{U}{kT}} + \frac{n_0}{3}\frac{\Delta U}{kT} v e^{-\frac{U}{kT}}. \qquad (VII.52)$$

Equation (VII.52) can be solved quite simply by assuming that $E = \text{const}$ ($E = F$) and, therefore, $\Delta U = \text{const}$. We use the definition [Eq. (VII.49)]

$$\tau = \frac{e^{U/kT}}{2v}$$

and solve Eq. (VII.52) with the imposed restrictions to obtain

$$\Delta n = ce^{-t/\tau} + \frac{n_0 \Delta U}{6kT},$$

where c is an arbitrary constant equal to $-n_0\Delta U/6kT$ provided that $\Delta n = 0$ at $t = 0$.

Thus,

$$\Delta n = \frac{n_0 \Delta U}{6kT}(1 - e^{-t/\tau}) = \frac{n_0 e E \delta}{12kT}(1 - e^{-t/\tau}). \qquad (VII.53)$$

The quantity τ can be called the relaxation time of weakly bound ions. It represents the rate of increase of Δn with time after the application of an electric field. Under steady-state conditions, i.e., at $t = \infty$, we have

$$\Delta n_\infty = \frac{n_0 \Delta U}{6kT} = \frac{n_0 e E \delta}{12kT}. \qquad (VII.54)$$

We recall that this expression is valid only when the work done by the field in the mean "free" path of an ion is small compared with the energy of thermal motion ($\Delta U \ll kT$).

Thus, the application of a field polarizes a dielectric containing weakly bound ions. Each ion transferred by the application of a field produces a dipole moment equal to $e\delta$. Therefore, the electric moment per unit volume is

$$P = \Delta n e \delta = \frac{n_0 e^2 \delta^2}{12kT} E. \qquad (VII.55)$$

Hence, the equivalent thermal ionic polarizability of an ion is

$$\eta_{it} = \frac{e^2\delta^2}{12kT}. \tag{VII.56}$$

The value of η_{it} decreases with increasing temperature since high temperatures impede the ordered motion of particles.

We note that the dipole moment ($\mu = e\delta$), due to a displaced ion in the thermal ionic polarization, is constant and independent of the applied field. It follows that the equivalent polarizability $\eta = \mu/E = e\delta/E$ is inversely proportional to the field; the polarization of a dielectric is approximately directly proportional to the field.

The thermal ionic polarization is usually observed in amorphous (glassy) and polycrystalline (ceramic) inorganic dielectrics. This type of polarization is of little importance in pure crystals. It is more important in crystals containing various impurities. These points will be taken up again in connection with the polarization processes in radiation-damaged ferroelectrics, and in the discussion in Chap. VIII of the electrical conductivity.

C. Interfacial and High-Voltage Polarization. The interfacial polarization is observed in heterogeneous dielectrics and is due to differences between the values of the electrical conductivity and permittivity of different parts (or microparticles) of a dielectric. When an electric field is applied to such a dielectric charges accumulate at the boundaries between different components and this is equivalent to polarization.

The interfacial polarization can be regarded also as a relaxational polarization because a definite time is required for its establishment and destruction and, moreover, it generally obeys an exponential law. However, the interfacial polarization is quite different from the thermal orientational and thermal ionic polarizations because no rotation of dipoles and no transitions of weakly bound ions take place in the establishment of the interfacial polarization. The interfacial polarization depends strongly on the shape, concentration, and actual properties of the components of a heterogeneous dielectric. It is very difficult to discuss this type of polarization in general. The interfacial polarization is frequently encountered in multicomponent dielectric systems. Some properties of the interfacial polarization will be discussed later in connection with the decay of the current in heterogeneous dielectrics (§ 2 in Chap. VIII).

The high-voltage polarization represents macroscopic displacements of charged particles in solid dielectrics under the action of strong electric fields. Such displacements give rise to an ionic or electronic space charge, which is positive near the cathode and negative near the anode. For this reason it is also known as the space-charge polarization. The space charge gives rise to an electric field directed opposite to the applied field. This is equivalent, in the macroscopic sense, to the establishment of a polarization.

Processes associated with the establishment of the high-voltage (space-charge) polarization are of the relaxational type and are characterized by long relaxation times amounting to seconds, minutes, and sometimes even hours. When the external field is switched off, the space charge is gradually dispersed by the internal high-voltage polarization field and the polarization disappears.

This type of polarization can be observed in many crystals. It will be discussed in more detail in Chap. VIII.

§4. Polarization of Ferroelectrics

in Static and Pulsed Fields

Outside the ferroelectric range of temperatures, i.e., in the ordinary paraelectric modification, the electric polarization of ferroelectric crystals does not exhibit any special features. Such crystals may exhibit the electronic displacement polarization and, depending on the structure, the ionic displacement polarization as well as the thermal ionic polarization. The paraelectric modification is not polar and, therefore, it cannot exhibit the thermal orientational polarization (see Chap. IV).

The ferroelectric modification is characterized by the presence of domains (see Chap. IV). A single-domain crystal or a single domain by itself can, in general, exhibit electric polarization of the type encountered in ordinary dielectrics if an external field does not change the orientation of the polar direction

of such a crystal or domain. Crystals of ordinary linear pyroelectrics (tourmaline, saccharose, lithium sulfate, etc.) behave
similarly. The situation is different in the case of ferroelectrics
which are split into domains. In addition to ordinary types of polarization, we encounter here the polarization associated with domain orientation. The contribution of this polarization is always
much larger than the contributions of the other mechanisms. In
view of this, we shall consider mainly the polarization mechanisms associated with the domain orientation: increase in size
of domains of one orientation, disappearance of domains of other
orientations, nucleation of new domains, and their growth.

In general, the polarization of a ferroelectric is a fairly complex function of the electric field intensity, the duration of application of the field, and the temperature. In particular, the presence
of domains makes the field dependence of the polarization nonlinear and the value of the polarizability α at a constant temperature becomes a function of the field

$$\alpha = \alpha \, (E). \tag{VII.57}$$

Ferroelectrics are the most typical representatives of the class of
nonlinear dielectrics. The dielectric hysteresis of the polarization in alternating fields is associated with the domain structure.
This hysteresis indicates that the state of polarization of a ferroelectric depends on its previous history.

The mobility of ferroelectric domains, characterizing the
velocity of motion of a growing nucleus of a domain, depends on
the temperature of a crystal and the closeness of this temperature to the phase-transition point.

The degree of polarization of a ferroelectric depends not only on the field intensity but also on the duration of application of
the field to a crystal. Hence, it follows immediately that the polarizability of a ferroelectric depends, other conditions being equal,
on the frequency of the applied electric field.

Investigations show that the total reversal of the polarization of a ferroelectric (polarization reversal or switching) can be
achieved only when the external field is higher than a certain critical value which is known as the coercive field in the case of cyclic polarization reversal. The value of the critical field is re-

lated to the nucleation and motion (growth) of new domains. The
critical field is that field whose application ensures effective nucleation and motion (growth) of domains oriented along the field.
It follows also that the coercive field depends on the polarization
reversal (switching) frequency. That part of the process of polarization of a ferroelectric which is associated with the nucleation of new domains is random; it is characterized by a definite
value of the activation field (this field is a function of the state of
the surface of a crystal, its temperature, etc.) The polarization
in such cases is characterized by a definite relaxation time.

The extensive material available on the polarization of ferroelectric crystals prevents us from analyzing in detail all the relationships governing this process. We shall consider simply the
most important properties, which have been firmly established.

A. Static Fields. The duration of the static polarization processes (lasting up to several hours) makes it difficult to
investigate them by electric methods if only because of a finite
(although low) electrical conductivity of ferroelectric crystals. In
view of this, the greatest interest lies in the static investigations
of the polarization carried out by optical methods in which observations are made of changes in the domain structure. The most
thorough optical investigations of this type type have been carried
out on the polarization of Rochelle salt, barium titanate, and several other crystals.

Some idea of the time dependence of the polarization reversal
in a Rochelle salt crystal subjected to various field intensities can
be gained from Fig. 165. The ordinate of this figure represents the

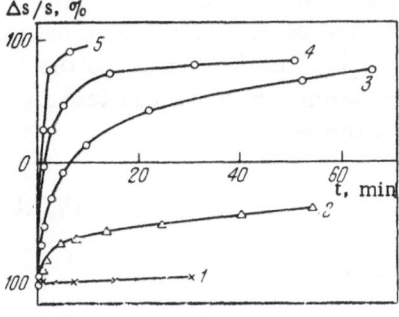

Fig. 165. Dependence of the domain
structure of Rochelle salt on time under
the influence of an electric field. The
initial state of the sample is assumed to
be that produced by the application of a
field of 400 V/cm for 30 min. The applied fields are (in V/cm): 1) 20; 2)
40; 3) 50; 4) 60; 5) 80 [15].

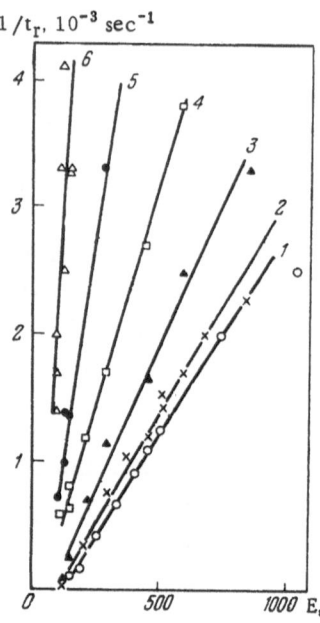

1/t_r, 10^{-3} sec^{-1}

Fig. 166. Dependence of the reciprocal of the polarization reversal time, $1/t_r$, of Rochelle salt on the field intensity at various temperatures (°C): 1) 8.8; 2) 10.7; 3) 5.5; 4) 17.7; 5) 20.4; 6) 23.4 [15].

quantity $\Delta s/s$ (Δs is the difference between the areas of domains of opposite orientation and s is the area of domains of both orientations), which represents the degree of polarization of a sample, originally polarized along one direction only. It is evident from Fig. 165 that, up to ∼ 20 V/cm, no polarization reversal takes place, no matter for how long the field is applied. A field of 40 V/cm alters the state of Rochelle salt but is incapable of producing complete polarization reversal. A field close to 50 V/cm can be regarded as the critical polarization reversal field. This field ensures practically total reversal in a time of the order of 1 h. An effective polarization reversal can be achieved in a shorter time if fields stronger than 50 V/cm are used. The nature of the dependence of the quantity $\frac{1}{2}(\Delta s/s + 1)$, which is equal to the fraction of the reversed volume of a sample in near-critical fields, can be described by an expression of the type

$$\frac{1}{2}\left(\frac{\Delta s}{s} + 1\right) = 1 - e^{-t/\tau}, \qquad (VII.58)$$

where the relaxation time τ is about 25 min.

Measurements of the time t_r, necessary for the total reversal of the original polarization, carried out in a wide range of field intensities (150-1000 V/cm), yield curves similar to those shown in Fig. 165, which indicate that in supercritical fields the dependence $1/t_r = f(E)$ for Rochelle salt is linear and the slopes of the lines depend on temperature (Fig. 166). It is evident from this figure that extrapolation of the straight lines to the reversal time $t_r \to \infty$ gives a critical field $E_{cr} \approx 60$ V/cm, which is close to the value of E_{cr} for the sample whose properties are illustrated in Fig. 165. (The small difference between the values of E_{cr} for samples of Figs. 165 and 166 is unavoidable because of the inhomogeneity of Rochelle salt crystals.) It is worth mentioning that E_{cr} of Rochelle salt decreases slowly on approach to the upper Curie point.

The static polarization processes in Rochelle salt subjected to weak fields can be investigated quite simply by direct observation of changes in the domain structure by time-exposure photography or by cinematographic methods. Investigations carried out by these methods show that in subcritical fields (curve 1 in Fig. 165) a slight increase of the polarization with time is mainly due to the appearance of a small number of domains which have not grown fully in the preceding polarization stage. In an electric field the ab and ac domains of Rochelle salt behave in the same manner, although the ab domains are encountered more frequently, the walls between them are more stable, and they have lower energies.

The rate of nucleation of domains is found to depend exponentially on the field intensity and is described by the relationship

$$\frac{\partial n}{\partial t} = ce^{-\alpha/E}, \tag{VII.59}$$

where n is the number of domains, c is a constant, and α is the activation field.

The random nature of the polarization in near-critical fields [see Eq. (VII.58)], in which many new domains are nucleated, suggests that the dominant factor is the fraction of the volume of a crystal in which the polarization is reversed by the nucleation of new domains. The polarization reversal time in such fields should depend exponentially on the field intensity.

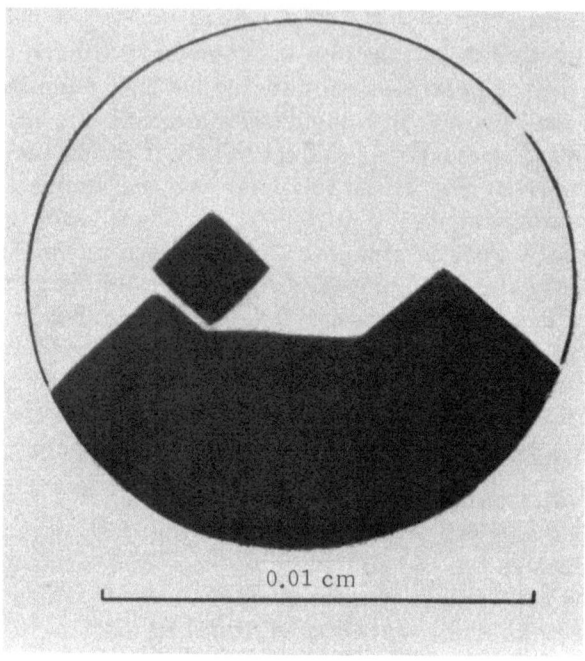

0.01 cm

Fig. 167. Photomicrograph of domains in a BaTiO$_3$
single crystal partly reversed by a field of 300 V/cm
(the polarization-reversed regions are shown dark).

The velocity of sideways motion v of domain walls in Rochelle salt subjected to supercritical fields is found experimentally to be proportional to the field:

$$v = \mu^* (E - E_{cr}), \qquad\qquad (VII.60)$$

where μ^* is the mobility of the sideways motion of the domains;
E_{cr} is the critical field. Estimates of E_{cr} and μ^* for a Rochelle
salt sample at 11°C give values of 80 V/cm and 1.36×10^{-3} cm$^2 \cdot$
V$^{-1} \cdot$sec^{-1}, respectively.† In a field of 200 V/cm at the same temperature of 11°C the domain wall velocity v is 0.14 cm/sec [62].
We have already mentioned that in supercritical fields the reciprocal of the polarization reversal time $1/t_r$ is proportional to

_____
†This value of E_{cr} corresponds to the motion of domains in a crystal which is in a
  multidomain state, while the value of E_{cr} determined from Fig. 166 corresponds to
  the single-domain state. In general, these fields should be different.

the field (Fig. 166). Comparing this result with the experimentally established relationship (VII.60), we may conclude that in strong
fields the polarization reversal time is governed primarily by the
time of growth of a new domain and not by its nucleation time, because in such fields the nucleation time is relatively short. If we
assume that a domain nucleus is in the form of a wedge and that
the motion of this wedge takes place in some viscous medium, we
find that the velocity of the sideways motion of a domain (its broad-
ening) is proportional to the field [Eq. (VII.60)].

The main experimental data on the mechanism of polarization reversal in $BaTiO_3$ crystals subjected to static fields were
obtained by Miller and Savage [61a, 61b, 61c, 61d]. They used
liquid electrodes (a solution of lithium chloride). After application of an electric field and partial switching of a crystal, its surface is etched and as a result the domain pattern is revealed. It is
found that polarization reversal in $BaTiO_3$ crystals subjected to
weak fields can take place by the sideways growth of a small number of domains (Fig. 167).

Investigations have also been made of the dependence of the
sideways velocity of a domain wall on the applied field. In this
method a small notch is made deliberately on the surface of a crystal and it serves as a nucleus of a new domain which grows during
polarization reversal. Repeated application of the field and etching of the crystal can be used to determine the growth of a domain
with increasing field intensity (Fig. 168). It has been established that
in fields of a few hundred volts per centimeter (these are weak

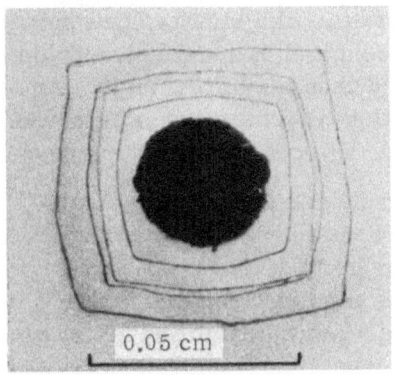

Fig. 168. Photomicrograph of a
$BaTiO_3$ crystal with a small pit, which
was partly reversed (switched) and
etched four times.

0.05 cm

fields in the case of BaTiO$_3$) the motion of a 180° wall can be described by an equation of the type

$$v = v_\infty e^{-\delta/E} \qquad\qquad\qquad (VII.61)$$

in a range of velocities covering three orders of magnitude. The
value of v$_\infty$ is about 10 cm/sec and the constant δ amounts to a
few thousand volts per centimeter. The velocity of a domain wall
in weak fields can vary from about 0.1 to 0.01 cm/sec. This velocity depends on the temperature: when the temperature is increased
the velocity increases mainly because of a decrease of the value
of δ.

Investigations of BaTiO$_3$ crystals have also shown that the
sideways motion of a wall in a field is fairly smooth and the value
of the critical field E$_{cr}$ may be assumed to be zero, i.e., any field
can displace a wall. In strong fields the wall velocity reaches
100 cm/sec. Moreover, in strong fields v$_\infty$ and δ are completely
independent of the field. The orientation of the walls of a growing domain also depends on the field. In weak fields these walls
are planes of the (110) type. In stronger fields we observe the
appearance of walls of the (100) type, and these are the only walls
present in still stronger fields. Experiments with metal electrodes show that the field dependence of the velocity of domain
walls in BaTiO$_3$ is exponential [Eq. (VII.61)] and the activation
field δ decreases with increasing temperature.

Comparison of the data on the motion of domain walls in
BaTiO$_3$ and Rochelle salt shows that the mechanisms of motion in these
two materials are different. This applies particularly to the absence
of critical fields for the sideways motion of the domain walls in
BaTiO$_3$. Moreover, the exponential dependence of the domain wall
velocity on the field intensity suggests that the sideways motion of
domain walls in BaTiO$_3$ is only apparent and that, in fact, new domains appear along existing walls and then they grow right up to
the surface of a crystal; this gives the appearance of a sideways
motion of the walls.

The sideways motion of 180° walls has been observed also
(using the method of charged powders) in crystals of triglycine
sulfate, but this motion is not as clear as in BaTiO$_3$. The difficulty is that, in practice, we cannot study the polarization reversal in only one domain. It is evident that conditions favorable

for the nucleation of many domains should be established during crystal growth.

B. Pulsed Fields. There have been very many investigations of polarization in pulsed electric fields because they yield information not only on physical processes in ferroelectrics but also on those characteristics which are important in possible applications of ferroelectrics as bistable elements (memory elements in digital computers, elements in logic circuits, etc.).

Very important information on the polarization processes in ferroelectrics can be obtained by varying (within wide limits) the amplitude, duration, and the sequence of the voltage pulses applied to ferroelectric crystals. This information is obtained from measurements of the current-pulse parameters and of the kinetics of polarization reversal (switching) using electrical and sometimes optical methods. The basic circuit used in such measurements is shown in Fig. 169a. The same figure shows the time dependence of the current through a polarized ferroelectric crystal subjected to a voltage pulse of opposite polarity (which reverses the polarization, Fig. 169b), and of the same polarity as the crystal (Fig. 169c). In electronic pulse circuits the first of these signals can be regarded arbitrarily as unity (1) and the second as zero (0).

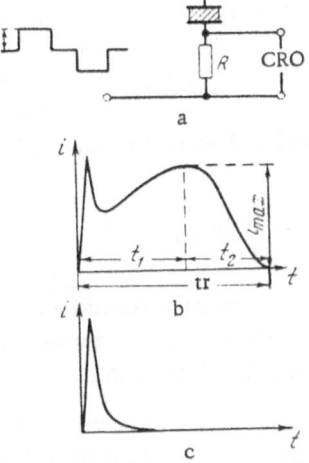

Fig. 169. Pulsed polarization reversal in ferroelectrics: a) measuring circuit for determination of the polarization reversal time t_r and the polarization reversal current i, as a function of the applied field E; b) polarization reversal current as a function of time during application of the first pulse; c) polarization reversal current during the application of a second pulse in the same direction.

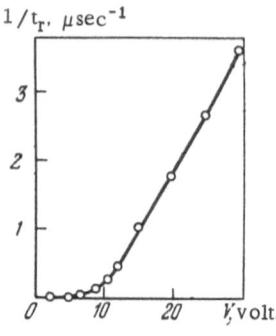

Fig. 170. Dependence of the reciprocal of the polarization reversal time of BaTiO$_3$ on the applied voltage [59].

In physical experiments it is usual to employ symmetrical bipolar voltage pulses, which produce polarization reversal for both polarities. The ferroelectric crystal under investigation is thus subjected to a sequence of fairly rapid polarization reversals, which distinguishes this condition from those obtaining in static tests in which polarization reversal can be regarded as a once-only process. The first pulse investigations of barium titanate were carried out by Merz [12, 58, 60]. The results of these investigations are still of great importance. The most interesting result obtained by Merz is the dependence of the total polarization reversal time t_r on the applied electric field. This dependence is shown in Fig. 170 for BaTiO$_3$. It is evident from this figure that the dependence of the reciprocal of the polarization reversal time $1/t_r$ on the field intensity is nonlinear in weak fields but it becomes linear in strong fields. In the weak-field region this dependence can be described by the relationship

$$\frac{1}{t_r} = \frac{1}{t_0} e^{-\alpha/E}, \qquad\qquad (VII.62)$$

where t_0 is a constant and α is the activation field. In strong fields the dependence is of the type

$$\frac{1}{t_r} = \frac{v}{d} = \frac{\mu}{d}(E - E_0) = \frac{\mu}{d^2}(V - V_0), \qquad\qquad (VII.63)$$

where v is the "forward" velocity of growth of a domain nucleus; μ is the domain mobility; d is the thickness of a sample; E_0 is the coecive field; $(V - V_0)$ is the potential difference.

Optical investigations, carried out in parallel with electrical measurements, have demonstrated that the application of a voltage pulse of polarity opposite to that of a polarized ferroelectric produces many small nuclei on the surface of a crystal and the polarization of these nuclei coincides with the direction of the pulse field. The nucleation of these new domains is followed by their growth. These investigations show also that the domains in $BaTiO_3$ grow mainly in the forward direction (across a crystal) but their sideways growth, as represented by the motion of the 180° walls, is very slow.

Comparing the results of the electrical and optical investigations, we may conclude that the polarization-reversal time of $BaTiO_3$ crystals is governed by two factors: the nucleation time t_n (this is the time for the formation of all nuclei from the first to the last) and the growth time t_g of a single domain right through a crystal. Since the nucleation process obeys statistical laws, we may expect an exponential dependence of t_n on the field. The growth of a domain right through a crystal can only have a linear dependence on the field, because it is due to forces resembling friction in a viscous medium.

The following polarization reversal mechanism can be suggested on the basis of these observations. In weak fields domain nucleation is a slow process compared with domain growth; therefore, the nucleation time governs the duration of the whole process and its exponential dependence on the field. Conversely, in strong fields all the nuclei are formed in a very short time and the polarization reversal is governed by the domain growth time, which is a linear function of the field.

This mechanism is based on the assumption that two phenomena are important in the polarization reversal of $BaTiO_3$: domain nucleation and domain growth. It is assumed that domain growth takes place mainly in the forward direction, i.e., the true sideways motion of the domain walls is very slow. We note that the considerable sideways motion observed in the static polarization reversal in $BaTiO_3$ crystals does not contradict this conclusion because this motion is (as pointed already) the result of the successive appearance of new nuclei at the walls of old domains.

The absolute value of the domain wall mobility μ [Eq. (VII.63)] in $BaTiO_3$ crystals subjected to strong fields (> 10^4 V/cm) at room temperature is about 2.5 cm$^2 \cdot$V$^{-1} \cdot$sec^{-1}. This value of the mobility in fields of the order of 10^4 V/cm corresponds to a domain

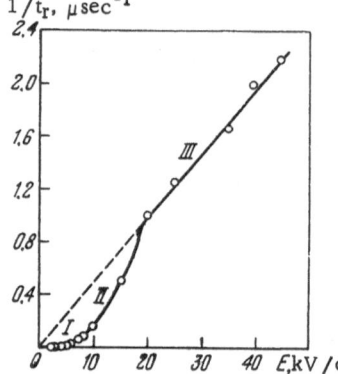

Fig. 171. Dependence of the reciprocal of the polarization reversal time of triglycine sulfate on the applied field [39].

growth velocity of about $(1-5) \times 10^4$ cm/sec. The domain velocity increases with increasing temperature.

Pulse measurements on triglycine sulfate crystals, carried out by Fatuzzo and Merz [39], confirm the principal ideas about the polarization reversal mechanism established for $BaTiO_3$ crystals. The dependence of $1/t_r$ on E for triglycine sulfate is shown in Fig. 171. It is evident from this figure that the dependence has an inflection point and that extrapolation of the linear part of the curve from this point passes through the origin. Thus, we can distinguish clearly the mechanisms of polarization reversal in triglycine sulfate due to domain nucleation (the curvilinear part of the dependence represented by I and II) and due to domain growth (the linear part of the dependence denoted by III). The domain mobility μ for the growth mechanism is estimated to be 2 cm$^2 \cdot$ V$^{-1} \cdot$ sec^{-1}.

Special investigations have been carried out to refine our knowledge of the polarization reversal mechanism. Thus, $BaTiO_3$ and triglycine sulfate crystals subjected to strong fields exhibit a quadratic dependence of the polarization reversal time t_r on the thickness of a crystal d (these measurements were carried out applying the same field to crystals of different thicknesses). This dependence is in agreement with Eq. (VII.63), based on the assumption that the polarization reversal time in strong fields is governed mainly by the domain growth mechanism. However, if we assume that the dominant process is the sideways motion of the walls (i.e., if we assume that the sideways growth associated with nucleation

is faster than the forward motion), the relationship describing
such a process [similar to Eq. (VII.63)] should be in the form of a
linear dependence of t_r on d. Thus, the experimental evidence
shows that the motion of a domain (or a nucleus) in the forward
direction is indeed slower than the sideways motion of the domain
walls in strong fields.

Important information on the polarization reversal mechanism can also be obtained by investigating the shape of the current pulses produced by the polarization reversal (Fig. 169). The
shape of these current pulses depends on the field intensity. We
can show that this shape is a function of the ratio of the time necessary for the nucleation of domains to the time of growth of new
domain nuclei. We shall denote the rise time of the current pulse
by t_1 (the first peak in the dependence of the current on the time
in Fig. 169b is due to charging of the "geometrical" capacitance of
the crystal; it lasts only for a short time and is ignored in calculations). Experiments show that a pulse of the current flowing
through a triglycine sulfate crystal in weak fields is strongly asymmetrical because it has a short rise time (t_1) and a long decay time
(t_2). Newly formed domains rapidly penetrate a crystal. The process of domain growth corresponds to the rising part of the current pulse. However, the total polarization reversal time t_r is, in
this case, governed mainly by the slower process of domain nucleation, corresponding to the falling part of the current pulse
$(t_2 > t_1)$. When a strong electric field is applied to such a crystal
the rate of formation of nuclei becomes so high that the polarization reversal time t_r is now governed mainly by the domain mobility and the current pulse becomes more symmetrical $(t_2 \approx t_1)$. Thus,
the mechanisms of domain nucleation and growth are quite separate in triglycine sulfate crystals: in weak fields the polarization reversal time is governed by the domain nucleation time, but
in strong fields it is governed by the domain growth time.

Reversal of the polarization of triglycine sulfate crystals by
pulses of shorter duration than the polarization reversal time
gives different results in strong and weak fields. For example, in
weak fields, the polarization reversal (switching) cannot be completed if the duration of the polarizing pulse is shorter than the domain growth time t_g even if a train consisting of very many pulses
is applied. Freshly nucleated domains do not have sufficient time

to grow right through a crystal and when the field is removed the domains are reduced to wedges and disappear completely.

A study of the shape of current pulses in $BaTiO_3$ crystals shows that in these crystals the domain nucleation mechanism is slower both in strong and weak fields. Results opposite to those for $BaTiO_3$ are obtained for guanidine aluminum sulfate. In this crystal, even in very weak fields, the slower mechanism is that of the growth of domains right through a crystal. A theoretical analysis of the polarization-reversal-pulse shape shows that their asymmetry, i.e., the inequality of the pulse rise and decay times, is a manifestation of the interaction between domain nuclei. Among the investigated crystals, the stronger interaction is exhibited by nuclei in $BaTiO_3$ and the weakest interaction by nuclei in guanidine aluminum sulfate. This is correlated with the highest and lowest values of the spontaneous polarization in $BaTiO_3$ and guanidine aluminum sulfate, respectively.

A two-stage polarization reversal mechanism (domain nucleation and their subsequent growth) can be investigated also by special thermal experiments. By varying the temperature of the surface and volume of a sample at different rates we can determine the influence of the temperature on the nucleation of new domains, the mobility of domain walls μ, and the activation field α.

In one of such experiments the surface of a triglycine sulfate plate is heated for a short time using an infrared radiation source. The duration of heating is sufficient to establish a temperature difference between the surface of a sample (where domain nuclei are formed) and its interior. In weak fields this treatment produces a small change in the duration of the rise time of the polarization reversal pulse t_1 and a strong reduction of the duration of the total polarization reversal time t_r and, consequently, of the decay time t_2. This means that the conditions for domain nucleation are affected by the temperature of the surface of a crystal. Surface heating accelerates the formation of domain nuclei, but the rate of growth of domains, governed by the temperature of the interior of a crystal, is not affected.

The process of polarization reversal in strong fields can be accelerated by increasing the temperature of the interior of a crystal. Uniform heating makes the polarization reversal current pulse shorter and larger in amplitude. Such heating reduces the

rise time t_1 (increases the domain wall mobility μ) as well as the
total reversal time t_r (reduces the activation field α). During the
subsequent cooling the surface of a crystal cools more rapidly than
its interior. Therefore, we find that, initially, the pulse decay
time t_2 increases and then (as the whole sample is cooled) the value
of t_1 also increases. Finally, the pulse recovers its initial shape.
A significant change in the polarization reversal time of guanidine
aluminum sulfate crystals can be obtainined only after prolonged
heating of the surface, which ensures that the whole crystal is
heated. This means that the growth of domain nuclei across a
quanidine aluminum sulfate crystal is slower than the nucleation
process. Weak interaction between fully grown domains and nuclei in this material is responsible for the very fast nucleation.

Investigations of the processes of polarization reversal in
Rochelle salt crystals subjected to pulsed fields yield dependences
$1/t_r = f(E)$, similar to those for BaTiO$_3$ (Fig. 172; see also Fig.
166). The value of the mobility of domain walls in Rochelle salt
is considerably higher than the corresponding mobilities in BaTiO$_3$
and triglycine sulfate: the mobility in Rochelle salt reaches about
100 cm$^2 \cdot$V$^{-1} \cdot$sec^{-1} in the middle of the ferroelectric range of temperatures. The mobility μ increases on approach to both Curie
points of Rochelle salt and near the lower Curie point the mobility is higher than at the upper point. The maximum value of the
activation field α in Eq. (VII.62) for Rochelle salt is about 650
V/cm at about 10°C. The activation field is considerably lower

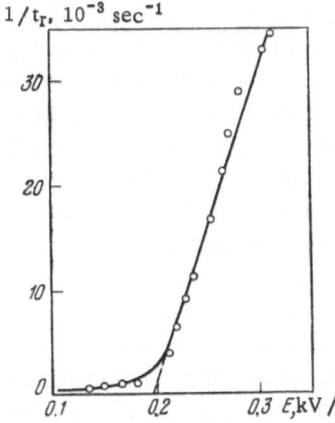

Fig. 172. Dependence of the reciprocal of the polarization reversal time $1/t_r$ of Rochelle salt on
the field [75].

near the phase-transition points: it is less than 100 V/cm near the lower Curie point and about 500 V/cm near the upper point.

The mechanism of pulsed polarization reversal in guanidine aluminum sulfate is similar to that in $BaTiO_3$. In weak fields guanidine aluminum sulfate exhibits an exponential dependence of the polarization reversal time on the field amplitude [Eq. (VII.62)]. The activation field α amounts to about 4.1×10^3 V/cm, i.e., it is of the same order as for the tetragonal modification of $BaTiO_3$. Guanidine aluminum sulfate does not have a critical field and the polarization can be reversed by a very weak field provided it is applied for a sufficiently long time. In strong fields the polarization reversal time is a linear function of the field intensity [Eq. (VII.63)] and the domain mobility in guanidine aluminum sulfate is $0.02-0.08$ $cm^2 \cdot V^{-1} \cdot sec^{-1}$, which is considerably lower than the mobility in $BaTiO_3$ (2.5 $cm^2 \cdot V^{-1} \cdot sec^{-1}$).

Analysis of the dependences $1/t_r = f(E)$, shown in Figs. 170–172, yields some information on the nature of the coercive field E_0, activation field α, and critical field (the minimum field required for polarization reversal of a crystal after a long time). It is evident from Figs. 170 and 172 that the coercive field E_0, i.e., the minimum field required to set a domain nucleus in motion, is finite both for $BaTiO_3$ and for Rochelle salt, amounting to 4.2×10^3 and 0.2×10^3 V/cm, respectively. The values of the activation field α for these two crystals are close to the corresponding values of the coercive field. The coercive field E_0 of triglycine sulfate is zero (the relevant curve extrapolates to the origin of the coordinates). This means that any domain nucleus in triglycine sulfate grows right through a crystal under the action of even a very weak field (provided this field is applied for a sufficiently long time). The activation field for triglycine sulfate at 30°C is about 4×10^3 V/cm. The higher values of the coercive field exhibited by $BaTiO_3$ and Rochelle salt (the coercive field of triglycine sulfate is equal to zero) can be explained as follows: the formation of nuclei and their motion in $BaTiO_3$ and in Rochelle salt are associated, in view of the symmetry of these crystals, with changes in the lattice structure which give rise to mechanical stresses, but the corresponding processes in triglycine sulfate crystals produce no such stresses. The finite coercive field E_0 for the growth of domains in $BaTiO_3$ is in apparent contradiction with our conclusion that the coercive field for the sideways motion of do-

main walls is zero since such motion is regarded as a result of
the growth of nuclei. This contradiction can be resolved by ob-
serving that the process of polarization reversal in BaTiO$_3$ involves not only the formation of the antiparallel (180°) domains
but also of the 90° domains.

Figures 170–172 show also that in all three crystals any
electric field, no matter how weak, is capable of reversing the polarization provided it is applied for a sufficiently long time. This
means that, within the limits of the experimental error, all the
curves in these figures do not intersect the field axis (the abscissa).

The field dependence of the polarization reversal time of
some crystals differs from the dependences shown in Figs. 170–
172 and described by Eqs. (VII.62) and (VII.63). It has been found
that, in the case of BaTiO$_3$ subjected to very strong fields, or in
the case of thiourea, lithium trihydrogen selenite LiH$_3$(SeO$_3$)$_2$, and
tetramethylammonium trichloromercurate, this dependence is of
the form

$$t_r = kE^{-n}, \tag{VII.64}$$

where k is a constant. In the case of thiourea and of BaTiO$_3$ in
very strong fields, we find that the power exponent is n $\approx$ 3/2.
The power exponent for LiH$_3$(SeO$_3$)$_2$ is approximately 5/2. The
value of a n for tetramethylammonium trichloromercurate is even
higher; moreover, this value depends quite strongly on the temperature, decreasing from about 7 at 20°C to about 3 at 200°C. A
power polarization reversal law with a small exponent (3/2, 5/2)
can be explained to some extent by assuming the occurrence not
only of domain nucleation and growth but also of sideways wall
motion.

Detailed investigations have been carried out on LiH$_3$(SeO$_3$)$_2$
crystals, in which domain nucleation processes are slower than domain growth. These crystals, like barium titanate, may be switched
by a large number of short duration pulses, the duration being less
than the polarization reversal time but greater than a certain critical value t$_{cr}$.

The spontaneous polarization of LiH$_3$(SeO$_3$)$_2$ is high and the
interaction between the domain nuclei is, as in BaTiO$_3$, quite strong.
This is why the nucleation of new domains is slow in both crystals. Finally, we must point out that the observation that the nu-

cleation or growth processes are slower or faster in weak or strong fields does not indicate at all that the function $1/t_r = f(E)$ is of the same form in all fields. Thus, we have mentioned already that domain nucleation in $BaTiO_3$ crystals subjected to weak and strong fields is slower than domain growth and, therefore, the growth determines the duration of polarization reversal but not the nature of the dependence $1/t_r = f(E)$. The exponential form of the dependence, observed for $BaTiO_3$ in weak fields, indicates that in such fields the variable component of the duration of the switching current pulse is the component representing domain nucleation. Conversely, in strong fields, the variable component of the duration of the switching current pulse is that component which represents domain growth, in spite of the fact that the total polarization reversal time is governed primarily by the nucleation time.

§5. Polarization of Ferroelectrics in Weak Sinusoidal Fields

The most typical experiments carried out on ferroelectrics are those concerned with polarization processes in sinusoidal (alternating) fields. A considerable number of such investigations has been carried out in weak fields (of the order of a few V/cm) for the purpose of measuring the permittivity ε. Since the appearance of the ferroelectric modification is associated with the appearance of the spontaneous polarization and domain structure, which govern the nature of polarization of a crystal in external fields, measurements of the temperature dependences of the weak-field permittivity are used very extensively in the detection and studies of ferroelectric phase transitions.

To obtain more or less complete data on the polarizability of ferroelectrics it is essential to investigate not only the dependence of the permittivity on the electric-field amplitude (such dependences give information on the nonlinearity of a ferroelectric) but also the frequency dependence of the permittivity. Polarization processes in alternating fields can also be compared with changes in the domain structure in ferroelectrics.

A. Permittivity ε and Polarizability α of Ferroelectrics. A typical field dependence of the polarization of a ferroelectric, characterizing its nonlinearity [Eq. (VII.57)] is shown in Fig. 173. It is evident from this figure that the differ-

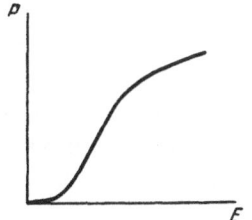

Fig. 173. Nonlinear dependence of the polarization on the field.

ential polarizability α, represented by the tangent of the slope angle at a given point on the curve, has different values at different field intensities. The polarizability in a weak field is characterized by the slope of the P(E) curve near the origin of the coordinates, i.e., in zero field:

$$\alpha_i = \left(\frac{\partial P}{\partial E}\right)_{E=0}. \tag{VII.65}$$

The permittivity of a crystal in weak fields ε_i is given by one of the relationships in Eq. (VII.1):*

$$\varepsilon_i = 1 + 4\pi\alpha_i.$$

In some cases (particularly near the phase transition temperatures) the permittivity of ferroelectrics is $\varepsilon_i \gg 1$ and, therefore, we can rewrite the last expression in the form of an approximate equality

$$\varepsilon_i \cong 4\pi\alpha_i. \tag{VII.66}$$

Since we shall now speak only of the values of ε and α in weak fields we shall drop the subscript "i."

Polarization processes in dielectrics are, in general, governed not only by the crystal structure, nature of chemical bonds, etc., but they are also influenced by the ability of a crystal to deform (or vibrate) in an electric field, i.e., these processes depend on the mechanical state of a crystal. The deformation of a dielectric in an electric field may be due to electrostriction or piezoelectric effects. The fraction of the electric energy dissipated in mechanical vibrations due to the piezoelectric effect is

*The permittivity in weak fields is frequently called the initial permittivity and denoted by the subscript "i."

represented by the electromechanical coupling coefficient k (see Chap. IX). In view of this, we must distinguish, in the case of piezoelectric crystals, the permittivity of a free crystal ε^t (i.e., a crystal which can deform freely and in which stresses t are equal to zero) and a clamped crystal ε^r (i.e., a crystal which has zero strain). The relationship between ε^t, ε^r, and k is of the form

$$\frac{\varepsilon^t}{\varepsilon^r} = \frac{1}{1 - k^2}. \tag{VII.67}$$

It follows from the above relationship that if $k = 0$, $\varepsilon^t = \varepsilon^r$, but when $k \neq 0$ the permittivity of a free crystal ε^t is higher than the permittivity of a clamped crystal, which is to be expected, because polarization processes are not impeded by mechanical effects in a free crystal.

Investigations show that the electromechanical coupling coefficient of linear piezoelectrics is small $(k \approx 0.1)$ and ε^t differs little from ε^r. On the other hand, the electromechanical coupling coefficient of ferroelectrics is high $(k \approx 0.6-0.7)$, which gives rise to a large difference between ε^t and ε^r.

Ferroelectrics may exhibit also fairly large electrostrictive strains. Therefore, in some cases, when an electric field does not produce piezoelectric strains in a ferroelectric, the values of ε^t and ε^r may differ considerably because of electrostriction. This is typical of ferroelectrics which are not piezoelectric in the paraelectric modification (oxygen-octahedral ferroelectrics, ferroelectric alums, triglycine sulfate, guanidine aluminum sulfate). These ferroelectrics do not exhibit the piezoelectric effect even after splitting into domains (see Chap. IV).

When the permittivity ε is measured in alternating fields up to frequencies corresponding to the piezoelectric or electrostrictive resonance frequency, we find that we obtain the value of ε^t. The same value is obtained also in static fields. Above the piezoelectric or electrostrictive resonance frequencies the deformation of a crystal is unable to follow oscillations of the field and we obtain the value of ε^r.

In addition to this difference between the values of ε^t and ε^r, we must distinguish the adiabatic (ε_S) and isothermal (ε_T)

permittivities. The subscripts S and T indicate constant entropy and temperature and they show that, in the first case, measurements are carried out under conditions which prevent heat exchange between neighboring elements of a sample (S = const), and, in the second case, the conditions are such that all parts of the sample have the same constant temperature (T = const). The first of these conditions is usually satisfied when the field used in measurements varies rapidly: individual parts of a vibrating sample (set in motion by the converse piezoelectric effect) do not have sufficient time to exchange heat. The isothermal values ε_T can be obtained when the field used in the measurements varies very slowly with time so that the temperature of the sample remains constant.

Let us consider an analytic relationship between the values of α_S^* and α_T^*, which are the reciprocals of the polarizabilities and, in the final analysis, govern the difference between the values of ε_S and ε_T. Let us assume that our electric field E is a function of the temperature T, the entropy S, and the polarization P. The value of α_S^* can then be represented in the form

$$\alpha_S^* = \left(\frac{\partial E}{\partial P}\right)_S = \left(\frac{\partial E}{\partial T}\right)_P \left(\frac{\partial T}{\partial P}\right)_S + \left(\frac{\partial E}{\partial P}\right)_T. \qquad \text{(VII.68)}$$

We shall now find the expression for $(\partial T / \partial P)_S$ using the condition that the free energy of a stress-free crystal is given by

$$dG = TdS + EdP.$$

It follows from this expression that $T = (\partial G / \partial S)_P$ and $E = (dG/dP)_S$ and, therefore, using Maxwell's relationships, we obtain

$$\left(\frac{\partial T}{\partial P}\right)_S = \left(\frac{\partial E}{\partial S}\right)_P.$$

This expression can be represented in the form

$$\left(\frac{\partial E}{\partial S}\right)_P = \frac{(\partial E/\partial T)_P}{(\partial S/\partial T)_P}$$

and substituted into Eq. (VII.68):

$$\left(\frac{\partial E}{\partial P}\right)_S = \left(\frac{\partial E}{\partial P}\right)_T + \frac{(\partial E/\partial T)_P^2}{(\partial S/\partial T)_P}. \qquad \text{(VII.69)}$$

Since $(\partial S / \partial T)_P = c_P / T$, where c_P is the specific heat at constant polarization P, and $(\partial E / \partial P)_T = \alpha_T^*$ by definition, we can write Eq. (VII.69) in the form

$$\alpha_S^* = \alpha_T^* + \frac{T}{c_P} \left(\frac{\partial E}{\partial T} \right)_P^2 . \tag{VII.70}$$

Using the relationship $P = \alpha E$, as well as the fact that the coefficient α^* in the expansion of the free energy of Eq. (VII.22) represents the reciprocal of the polarizability [see Eq. (VI.23)]

$$\alpha^* = \frac{1}{\alpha} , \tag{VII.71}$$

and, assuming that in the first approximation α^* is a linear function of the temperature [Eq. (VI.26)]:

$$\alpha^* = \frac{T - T_c}{C^*} \tag{VII.72}$$

(T_C is the Curie–Weiss temperature, C^* is a constant), we obtain

$$\left(\frac{\partial E}{\partial T} \right)_P^2 = \frac{1}{C^{*2}} \left(\frac{\partial P}{\partial T} \right)^2 .$$

Since, by definition $\partial P / \partial T = p$, the last expression can be represented in the form

$$\left(\frac{\partial E}{\partial T} \right)_P^2 = \frac{1}{C^{*2}} p^2,$$

where p is the pyroelectric coefficient, which can be used – in conjunction with Eq. (VII.68) – to deduce the explicit relationship between α_S^* and α_T^*:

$$\alpha_S^* = \alpha_T^* + \frac{p^2 T}{C^{*2} c_P} . \tag{VII.73}$$

It follows from the last relationship that the values of α_S^* and α_T^* are different only for pyroelectric crystals. This means that in the paraelectric nonpolar phase of ferroelectrics we have $\alpha_T^* = \alpha_S^*$. However, below the ferroelectric phase-transition temperature, when a crystal (or, more exactly, its domains) becomes pyroelectric, we find that the values of α_S^* and α_T^* become different and, therefore, ε_S and ε_T are also different.

   B. Curie – Weiss Law. The linear temperature dependence of the coefficient α^* (the reciprocal of the polarizability) in the expansion of the free energy G [this linear dependence is given by Eq. (VII.72)] and the approximate relationship of Eq. (VII.66) yields the following expression for the temperature dependence of the permittivity ε:

$$\varepsilon = \frac{C}{T - T_c} \tag{VII.74}$$

where $C = 4\pi C^*$ is the Curie constant, and the expression is known as the Curie–Weiss law. This law is satisfied quite well by many ferroelectrics in the paraelectric state, below the Curie temperature.

   The expression for the free energy G, given by Eq. (VI.22), can be used to obtain a relationship between the reciprocals of the polarizability α^* above and below the Curie point. The reciprocal of the polarizability for a free crystal α^{*t} above the Curie point is given by

$$\alpha^{*t} = \left(\frac{\partial^2 G}{\partial P^2}\right)_{P=0}. \tag{VII.75}$$

In the case of a phase transition of the second kind, we need retain only two terms in Eq. (VII.22) and, therefore, below the transition temperature we obtain

$$\alpha^{*t} + 3\beta^t P^2 = \left(\frac{\partial^2 G}{\partial P^2}\right). \tag{VII.76}$$

Moreover, using Eq. (VI.25), which applies to phase transitions of the second kind, we find that Eq. (VII.76) can be rewritten in the form

$$- 2\alpha^* = \left(\frac{\partial^2 G}{\partial P^2}\right). \tag{VII.77}$$

   Comparison of Eqs. (VII.75) and (VII.77) shows that the slope of the temperature dependence of α^* within the ferroelectric range of temperatures is twice as large as outside this range. A schematic representation of the temperature dependence $\alpha^*(T)$, based on Eqs. (VII.72), (VII.75), and (VII.77), is shown in Fig. 174a in the

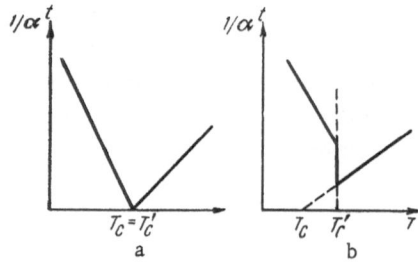

Fig. 174. Temperature dependence of the reciprocal of the polarizability $1/\alpha^t$ of ferroelectrics near phase transitions of the second (a) and first (b) kind.

region of the Curie temperature. It is evident from this figure that the permittivity of a free crystal ($\varepsilon^t = 1/\alpha^*$) is continuous but infinite at the transition point.

In the case of a phase transition of the first kind, it follows from Eqs. (VI.29) and (VI.30) that

$$P_s^2 = -\frac{3}{4}\frac{\beta_0}{\gamma_0},$$

$$\alpha^{*t} = \frac{3}{16}\frac{\beta_0^{t'}}{\gamma_0^t},$$

i.e., there is a discontinuity of the polarization and of the absolute value of the reciprocal of the polarizability α_f^* (and the coefficient $C*$) at the transition point. In the first approximation, the temperature dependence of the reciprocal of the polarizability can be assumed to be linear. A schematic representation of the function $\alpha^{*t}(T)$ is given in Fig. 174b. It is evident from this figure that there is a discontinuity of α^{*t} at a phase transition of the first kind and that the permittivity of a free crystal also has a discontinuity at this point, but does not become infinite.

The Curie–Weiss temperature,† denoted by T_C, is known to be identical with the true transition temperature (Curie point) T_c' at which the spontaneous polarization appears in ferroelectrics which exhibit a phase transition of the second kind. In the case of ferroelectrics with a phase transition of the first kind the Curie–Weiss temperature T_C is lower than the true transition temperature T_c' [see Eq. (VI.40)]. Usually, the difference ($T_c' - T_C$) amounts

---

† We must distinguish the concepts of the Curie–Weiss temperature and of the Curie temperature or point. We shall understand the Curie point to be the true transition temperature. This point coincides with the Curie–Weiss temperature only in the case of ferroelectrics exhibiting a phase transition of the second kind.

to about 10°. At frequencies higher than that of the mechanical resonance of a clamped crystal the permittivity of a ferroelectric
above the transition temperature obeys the Curie–Weiss law of
Eq. (VII.74) with the usual constant. However, in this case, the
Curie–Weiss temperature is lower than the corresponding temperature for a mechanically free crystal.

The absolute value of the Curie constant C of ferroelectrics
can be used to divide these materials into two groups. The first
group comprises those crystals (Table 40) whose Curie constant
C is of the order of 10^3 deg, and it includes ferroelectrics with orderable structure elements (Rochelle salt, KH_2PO_4, $NaNO_2$, etc.).
The second group comprises ferroelectrics of the oxygen–octahedral type: $BaTiO_3$, $PbNb_2O_6$, etc. The absolute value of the
Curie constant C for the second group is of the order of 10^5 deg. This
large difference between the values of C of the two groups of ferroelectrics provides an additional proof that there are two types
of change in the structure at phase transition points (this point is
discussed in connection with the structure of ferroelectrics, described in Chaps. II and III).

C. Low-Frequency Permittivity in Weak
Fields. We shall now consider briefly the temperature dependence of the weak-field permittivity at low frequencies (50–1000
cps).

The cubic (paraelectric) modification of barium titanate is
isotropic (from the polarization point of view) and has only one
value of ε. In the tetragonal modification we can distinguish the
permittivity along the spontaneous polarization direction (c axis),
denoted by ε_c, and the permittivity at right-angles to this direction (along the a and b axes), denoted by ε_a. These values of the

TABLE 40. Curie Constants

Substance	c, deg	Substance	c, deg
Rochelle salt	$2.2 \cdot 10^3$	$NaNO_2$	$5 \cdot 10^3$
KH_2PO_4	$3.3 \cdot 10^3$	$BaTiO_3$	$1.5 \cdot 10^5$
Colemanite	$0.5 \cdot 10^3$	$KNbO_3$	$2.0 \cdot 10^5$
TGS	$3.2 \cdot 10^3$	$Cd_2Nb_2O_7$	$1.0 \cdot 10^5$
Methylammonium	$1.0 \cdot 10^3$	$PbNb_2O_6$	$3.0 \cdot 10^5$
aluminum alum			

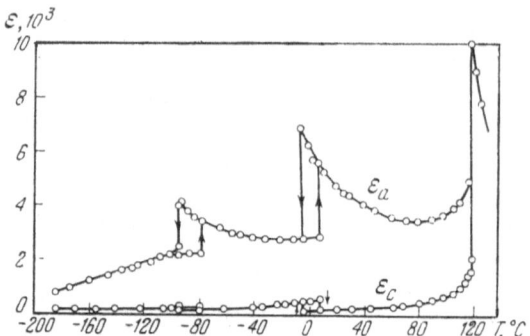

Fig. 175. Temperature dependence of the permittivity of a single-domain BaTiO₃ crystal, measured
along two mutually perpendicular directions, one
of which coincides with the spontaneous polarization direction in the tetragonal modification (c axis)
and the other is perpendicular to that direction (*a*
axis) [57].

permittivity can be measured using a single-domain sample or a
crystal in which the domain structure is deliberately reduced to
one domain. Measurements of all the components of the tensor
ε_{ij} in the orthorhombic and tetragonal modifications are difficult
and, in the majority of cases, the reported values are the average permittivities ε referred to the original cubic or tetragonal
axes.

The temperature dependence of the permittivity of BaTiO₃
is shown in Fig. 175. The permittivity ε of the nonpolar modification is described satisfactorily by the Curie–Weiss law $\varepsilon =
B + C/(T-T_c)$; the constant B is of the order of 10. Near the
transition temperature $\varepsilon \approx C/(T - T_c)$. The absolute values of
the Curie constant C, obtained by various workers, differ very
considerably. The Curie constant values are different because of
the different methods used to prepare the samples and to measure
their electrical conductivity. The most likely value of C for BaTiO₃
is listed in Table 40.

No appreciable deviations from the Curie–Weiss law are
exhibited by BaTiO₃, KNbO₃, or PbTiO₃. The Curie–Weiss temperature of these compounds is always lower than the true transition temperature T'_c. The value of ε at the transition point is finite
and falls with decreasing temperature. In the tetragonal modifica-

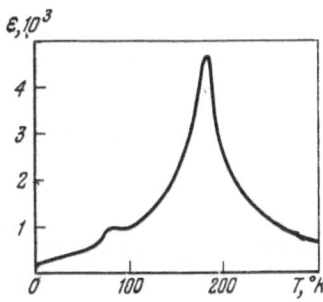

Fig. 176. Temperature dependence of the permittivity of a ceramic sample of $Cd_2Nb_2O_7$ [47].

tion the value of ε_c is much smaller than ε_a. This is because the appearance of the spontaneous polarization saturates the polarizability of these crystals along the c axis. The three ferroelectrics discussed in the present paragraph exhibit a thermal hysteresis of ε. The motion of domain walls in weak fields makes no appreciable contribution to the value of ε_c.

The temperature dependence of ε of cadnium niobate $(Cd_2Nb_2O_7)$, which has the pyrochlore structure, is shown in Fig. 176. According to this figure the phase transition in $Cd_2Nb_2O_7$ occurs at $-88°C$. Above this temperature the niobate obeys well the Curie–Weiss law with the Curie constant $C \approx 10^5$ deg.

The most interesting feature of the properties of lithium niobate $(LiNbO_3)$ and lithium tantalate $(LiTaO_3)$, which have the pseudoilmenite structure, is an increase in their spontaneous polarization P_S, permittivity ε_c, and coercive field E_c when the temperature is raised from room temperature (at this temperature

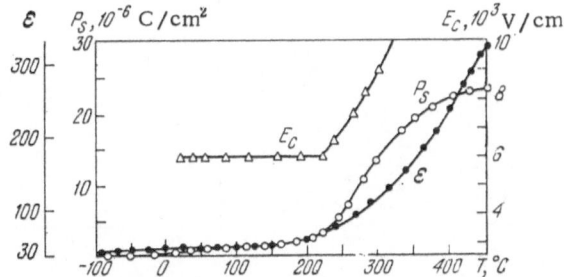

Fig. 177. Temperature dependences of the permittivity ε, spontaneous polarization P_S, and coercive field E_c of $LiTaO_3$ [56].

these compounds are ferroelectric), as shown in Fig. 177. Such behavior cannot be explained by an increase in the domain–wall mobility with increasing temperature since this would give rise to a reduction in the value of E_c (this point will be discussed later).

Triglycine sulfate exhibits a rapid rise of the permittivity, measured along the polarization axis b, when the transition temperature (49°C) is approached. The absolute value of ε_b reaches 10^4 at the transition point. The permittivity components ε_a and ε_c show no anomalies near the phase–transition temperature (Fig. 178a). It is evident from Fig. 178b that near the transition temperature the value of ε_b obeys the Curie–Weiss law with the Curie constant $C \approx 3.2 \times 10^3$ deg and the Curie–Weiss temperature which is identical with the transition temperature (+ 49°C).

As already mentioned, the tangent of the slope angle of the temperature dependence of $1/\varepsilon$ of ferroelectrics exhibiting a transition of the second kind (such as triglycine sulfate) should be different in the ferroelectric and paraelectric regions: the tangent should be twice as large in the ferroelectric region [see Eqs. (VII.75) and (VII.77)]. However, the ratio of the values of this tangent, obtained from Fig. 178b, is 2.7 instead of 2. This difference can be accounted for partly by the difference between the values of the adiabatic ε_S and isothermal ε_T permittivities. In fact, it follows from Eq. (VII.73) that in the paraelectric region (where the pyroelectric coefficient is $p = 0$) there is no difference

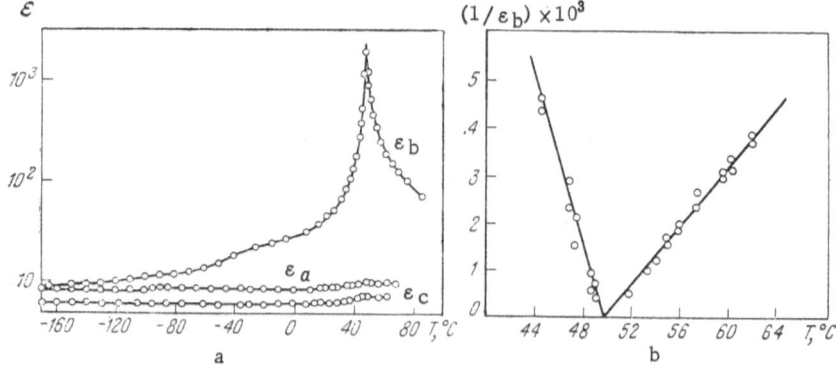

Fig. 178. Temperature dependences of the permittivity (a) and its reciprocal (b) for triglycine sulfate near the phase-transition temperature. The data for $\varepsilon = f(T)$ are taken from [46] and the data for $1/\varepsilon = f(T)$ are taken from [73].

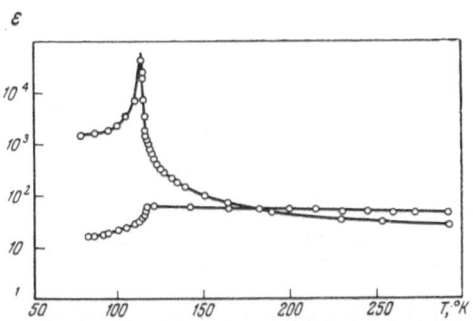

Fig. 179. Temperature dependence of the permittivity of KH_2PO_4 along the spontaneous polarization direction (ε_c) and at right-angles to this direction (ε_a). The measurements were carried out in a field of 200 V/cm at 800 cps [30].

between the values of ε_S and ε_T. Moreover, it follows from Eq. (VII.74) that, in this region,

$$\frac{1}{\varepsilon} = \frac{T - T_c}{C}$$

and

$$\frac{d}{dT}\left(\frac{1}{\varepsilon}\right)_{T > T_c} = \frac{1}{C}. \tag{VII.78}$$

In the ferroelectric region $\mathbf{p} \neq 0$ and, according to Eq. (VII.73), $\varepsilon_S \neq \varepsilon_T$. An analysis shows that this inequality gives a tangent ratio of 2.4 for the triglycine sulfate, which is quite close to the experimental value.

The temperature dependence of the permittivity of KH_2PO_4 is presented in Fig. 179: here, ε_c is the permittivity along the spontaneous polarization direction (the c axis of the tetragonal crystal), and in the paraelectric region the permittivity increases hyperbolically (in accordance with the Curie–Weiss law) on approach to the transition point. The value of the Curie constant of this crystal is $\sim 3.3 \times 10^3$ deg near the transition temperature. KH_2PO_4 crystals satisfy well the Curie–Weiss law right up to temperatures 50-60° higher than the transition temperature. Moreover, the Curie–

Weiss temperature T_c is identical with the phase-transition temperature.

KH$_2$PO$_4$ is a typical crystal exhibiting a phase transition of the second kind. The permittivity ε_c reaches very high values ($\sim 10^5$) at the Curie point, but it does not become infinite (as predicted by the theory of phase transitions of the second kind) because of inhomogeneities in the crystal structure, mechanical clamping, etc.

Below the transition temperature the value of ε_c of KH$_2$PO$_4$ falls very rapidly, but there is no discontinuity. The motion of domain walls, observed even in very weak fields, makes an appreciable contribution to the measured value of ε_c just below the temperature T_c. This contribution is manifested also by a considerable rise of the dielectric losses. In order to determine ε below the Curie point a crystal is converted to the single-domain state by the application of a constant field of about 500 V/cm. When a crystal is cooled below T_c in such a field, its structure reduces to a single domain. The permittivity ε_a, at right-angles to the spontaneous polarization direction, also changes in the vicinity of T_c but the change is not as great as that of ε_c. However, we must mention that it is necessary to distinguish ε_a and ε_c below the temperature T_c of KH$_2$PO$_4$.

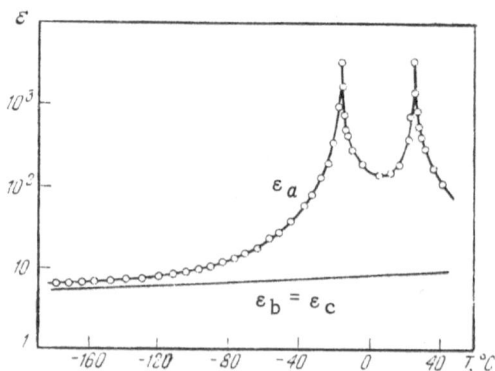

Fig. 180. Temperature dependence of the permittivity of Rochelle salt along the spontaneous polarization direction (ε_a, field intensity 4 V/cm) and along directions perpendicular to the spontaneous polarization (ε_b and ε_c, field intensity 50 V/cm) [44].

This prevents us from making final conclusions about the nature of the true changes in ε at right-angles to the c axis.

Rochelle salt exhibits a strong temperature dependence of ε only along the polarization axis a, i.e., only in the case of ε_a. At right-angles to the a axis the permittivity is small throughout the ferroelectric range of temperatures and it increases weakly with increasing temperature (Fig. 180).

It is evident from Fig. 180 that at both Curie points ($+24$ and $-18°C$) the value of ε_a has very sharp peaks at which it reaches values of a few thousands. No thermal hysteresis of ε_a is observed. Immediately above the upper Curie point ($+24°C$) the temperature dependence of ε_a obeys the Curie–Weiss law with a Curie constant $C = 2.24 \times 10^3$ deg. The Curie–Weiss temperature is approximately equal to the actual transition temperature. At higher temperatures (above 34°C) the Curie constant is $C = 1.71 \times 10^3$ deg. Below the lower Curie point ($-18°C$) the permittivity ε_a also obeys the Curie–Weiss law with $C = 1.18 \times 10^3$ deg. Below $-160°C$ the value of ε_a is about 7. The fall of ε_a in the interval between the Curie points represents a reduction in the domain-wall mobility μ (§4 in the present chapter). According to dielectric measurements, the phase transitions of Rochelle salt at both Curie points are of the second kind. The difference between the permittivities of free and clamped Rochelle salt crystals can be found from the experimentally determined difference:

$$\alpha_a^t - \alpha_a^r = 0,04.$$

The temperature dependence of the permittivity ε_c of sodium nitrite ($NaNO_2$) is shown in Fig. 181a. It is worth noting the thermal hysteresis ($\sim 3°C$) of the permittivity ε_c. Above the transition temperature ε_c obeys the Curie–Weiss law with $C = 5 \times 10^3$ deg. Below the transition temperature the value of ε_c falls at first rapidly and then very slowly, reaching about 6 at $-150°C$. At right-angles to the spontaneous polarization direction the low-temperature permittivities ε_a and ε_b are somewhat lower than ε_c, and they exhibit a slight anomaly near the transition temperature (Fig. 181b). It is evident from Fig. 181c that the Curie–Weiss law is obeyed. The ratio of the slopes of the straight lines in the ferroelectric region and above it is $\sim 8{:}1$. For phase transitions of the second kind this ratio should be $2{:}1$. This circumstance as

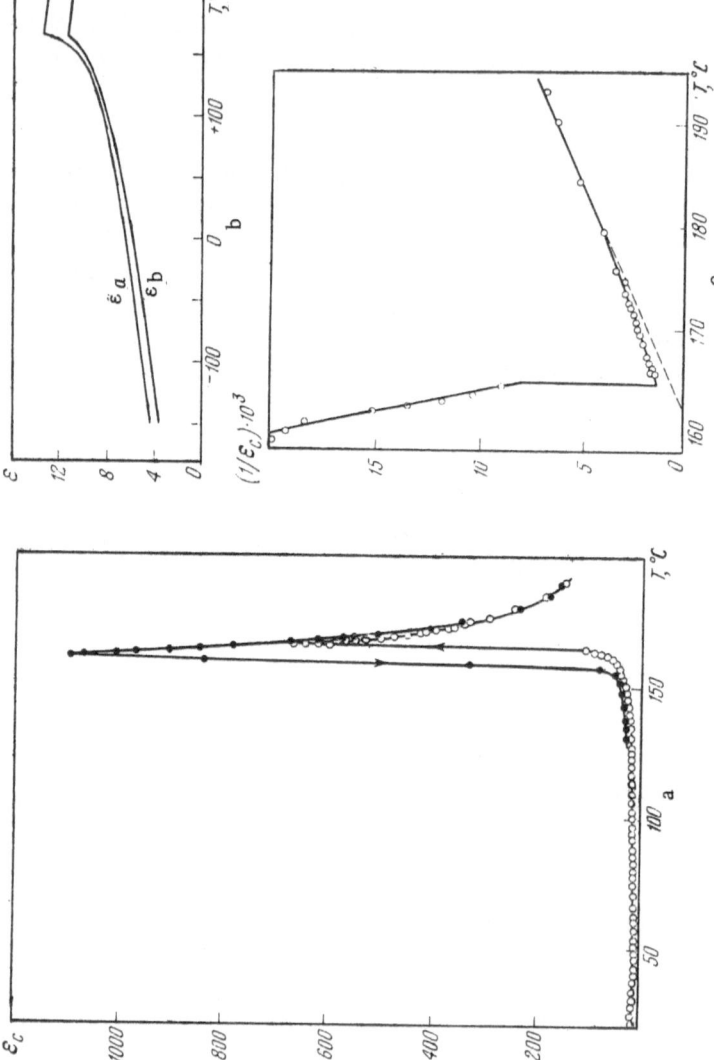

Fig. 181. Temperature dependences of the permittivity of NaNO₂: a) ε_c of a crystal grown from an aqueous solution and measured in a field of 100 kc [66]; b) ε_a and ε_b for a crystal grown from the melt [21]; c) $1/\varepsilon_c$ near a phase-transition point (c axis is the spontaneous polarization direction, and the results were obtained during heating) [66].

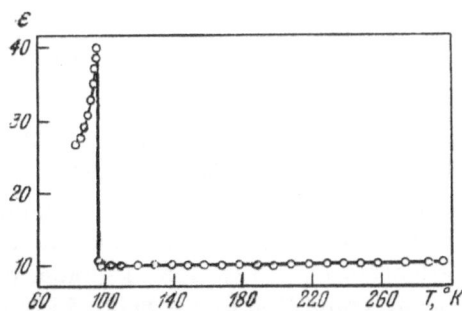

Fig. 182. Temperature dependence of the permittivity of a crystal of $Cd_2(NH_4)_2(SO_4)_3$, obtained by D. Eastman, F. Jona, and R. Pepinsky and reported in [10].

well as the thermal hysteresis of ε_c indicate that the phase transition observed in $NaNO_2$ at about $+161°C$ is of the first kind. However, other properties of $NaNO_2$ suggest that this phase transition has properties very similar to a transition of the second kind.

Unusual temperature dependences of the permittivities along the spontaneous-polarization direction, which does not satisfy the Curie–Weiss law above the phase-transition temperature, are reported for crystals of dicadmium diammonium sulfate $Cd_2(NH_4)_2(SO_4)_3$ and ammonium sulfate $(NH_4)_2SO_4$. The temperature dependences are similar for these two crystals. The temperature dependence of ε for one of them $[Cd_2(NH_4)_2(SO_4)_3]$ is shown in Fig. 182. The transition temperature of $(NH_4)_2SO_4$ is $-49.5°C$ and ε_c at this temperature is 160. However, the permittivity ε_b of am-

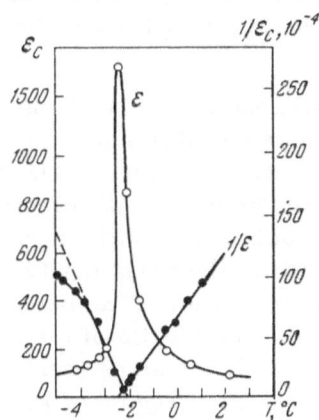

Fig. 183. Temperature dependence of the permittivity ε_c for NH_4HSO_4 [22].

monium fluoroberyllate NH_4BeF_4 obeys the Curie–Weiss law above the transition point ($-93°C$). The phase transitions in all these three crystals are of the first kind.

The permittivity of ammonium hydrogen sulfate NH_4HSO_4 obeys the Curie–Weiss law near $-3°C$ (Fig. 183). Moreover, the tangent of the slope angle of the temperature dependence of $1/\varepsilon_c$ in the ferroelectric region is twice as large as the corresponding tangent in the paraelectric region, which is typical of ferroelectrics exhibiting a phase transition of the second kind [see Eqs. (VII.75) and (VII.77)].

Guanidine aluminum sulfate has a small value of the permittivity, $\varepsilon_c = 6$ along the spontaneous polarization direction (the 6 axis); this value is obtained in weak fields at room temperature. This permittivity does not vary right up to 100°C, where dehydration processes begin. The permittivity at right-angles to the c axis is $\varepsilon_a \approx 5$.

D. HF and VHF Permittivity in Weak Fields. It is clear from the results reported in the preceding subsections that the permittivity of a ferroelectric may decrease with increasing frequency in the region where the field frequency is close to the mechanical resonance frequency of a given sample. At this frequency a sample changes from the mechanically free state to the clamped condition and this should reduce the value ε [Eq. (VII.67)]. This may be why many experimenters have reported a 30–40% fall of ε of $BaTiO_3$ crystals in the frequency range 1–1000 Mc (meter and decimeter wavelengths). In general, the frequency of the natural resonance of a crystal depends on its geometry and the nature of vibrations and it can be lower.

We have reported in §4 of the present chapter that the processes of polarization and polarization reversal in some crystals, which take place even in very weak fields, are associated with the growth of domains and the motion of domain walls. It follows that if there is a critical size of a domain nucleus (§8 of the present chapter), we should observe a dispersion of the permittivity at the frequency corresponding to its natural resonance. The permittivity dispersion may thus be associated with the inertia of the motion of the domain walls which may be attributed some inertial mass. We must also stress that the dispersion region of ε, associated as it is with the actual origin of the spontaneous polarization, is important in the case of ferroelectrics. The spontaneous polarization is due to the polarizability of the crystal structure (Chap. VI), the instability of lattice vibrations, etc. This means that the greatest interest in the dispersion of ε of ferroelectrics lies in the frequency range which corresponds, generally speaking, to the natural resonance frequencies of electrons, ions, and the crystal lattice.

Experimental investigations show that the dispersion of the permittivity of barium titanate, other perovskite ferroelectrics, and some of their solid solutions is observed in the range of frequencies corresponding to the centimeter wavelength band. Thus, according to Fousek [41], the midpoint of the dispersion region of ε of the c-domain crystals of $BaTiO_3$ lies at about 2000 Mc (λ = 15 cm) at room temperature; at 10,000 Mc (λ = 3 cm) $\varepsilon \approx$ 110-150, but at 1 Mc (λ = 3 $\times$ 10^4 cm) $\varepsilon \approx$ 2500-3500. Similar results have been reported also by other workers. However, we must mention that, according to Benedict and Durand [27], the value of ε of the a-domain crystals of $BaTiO_3$ retains its high value ($\cong$ 1500) right up to 2400 Mc ($\lambda \sim$ 1 cm). There have been reports that there is no dispersion of $BaTiO_3$ right up to 5000 Mc ($\lambda \sim$ 6 cm). Experimental investigations show that the dispersion of ε in the centimeter range is of the relaxational type. The dispersion frequency of perovskite dielectrics depends weakly on their compositions.

The dispersion of the permittivity ε, observed for $BaTiO_3$ crystals in the centimeter range of wavelengths, is explained in different ways by different authors. According to Kittel [52] and Sannikov [16], the dispersion is associated with the inertia of the domain walls, and it should occur at frequencies of 2 $\times$ 10^9 cps (according to Kittel) and 8 $\times$ 10^9 cps (according to Sannikov). These two authors conclude that the dispersion of ε of $BaTiO_3$ is indeed associated with the domain-wall motion. Without ignoring the role of the inertia of the domain walls in the relaxational processes in the centimeter range, Fousek [41] gives more prominence to processes associated with the relaxation of the dipolar component of the polarization in $BaTiO_3$. Applying the Mason–Matthias model to $BaTiO_3$ (in this model titanium ions occupy six positions separated by barriers of height ΔU), Fousek concludes that the relaxation frequency calculated on the basis of this model agrees with the experimentally observed frequency if ΔU is assumed to be

0.2 eV, which is in good accord with the values estimated by other workers.

The high values of the permittivity of $BaTiO_3$ in the centimeter range of wavelengths (even in the dispersion region) and the fall of the permittivity in static fields (§ 6 of the present chapter), as well as the absence of the permittivity dispersion above the Curie point, indicate that at these frequencies the domain processes play the dominant role. This means that the dispersion in the centimeter region can be explained without taking into account such processes as the thermal ionic polarization (i.e., models of the Mason –Matthias type), crystal lattice vibrations, etc. However, as pointed out in Chap. VI, the dynamic theory of lattices predicts that the dispersion of ε of $BaTiO_3$ should occur also in the millimeter range of wavelengths. Taking this into account, we may assume that the fall of ε of $BaTiO_3$ in the centimeter range of wavelengths can also be due to the onset of a strongly broadened resonance dispersion region of the transverse optical mode of the lattice vibrations.

A more or less full theory of the nature of the dispersion of ε of $BaTiO_3$ can be developed on the basis of measurements carried out on polycrystalline samples, the results of which are shown in Fig. 184a. The fall of the permittivity in the frequency range 10^9–10^{10} cps is due to the exclusion of domain processes at higher frequencies and the fall at frequencies $\sim 10^{12}$ cps is associated with natural vibrations of the lattice.

The frequency dependences of the weak-field permittivity and tan δ of triglycine sulfate are shown in Fig. 184b. Of greatest interest in this figure is the relaxation of the permittivity in the frequency range 10^5–10^7 cps; in this range ε falls from about 70 to 25. At higher frequencies the value of ε decreases slowly with increasing frequency. At 9.6 Mc ε amounts to 20.

The temperature dependence of ε at frequencies above the dispersion region is of the form characteristic of triglycine sulfate (Fig. 178). The Curie–Weiss law is obeyed throughout the investigated range of temperatures, and the Curie temperature is independent of the frequency. The Curie constant at 200 Mc is 740° in the ferroelectric region and 2500° above the transition temperature ($\sim$ 49°C).

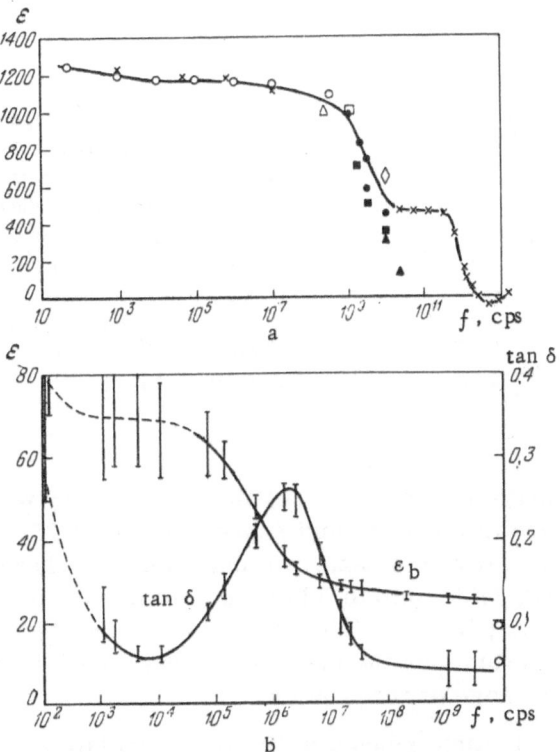

Fig. 184. Frequency dependences of dielectric properties
of ferroelectrics: a) dispersion of ε of ceramic $BaTiO_3$ in
a wide range of frequencies [13]; b) dispersion of ε_b and
tan δ of a single crystal of triglycine sulfate, recorded at
25.5°C [14a].

The reported relaxation of ε is observed only in the ferroelectric range of temperatures. This observation, as well as the
fact that the permittivity ε is independent of the alternating field
amplitude above the dispersion region (above 10^7 cps), show that
the relaxation of ε in triglycine sulfate is due to the exclusion of
the domain processes from the polarization. This is also confirmed by the rise of the coercive field of triglycine sulfate with
increasing frequency in the range from 10 to 10^5 cps and the absence of a dependence of ε on a bias field at frequencies above
the relaxation region. Thus, the domain walls in triglycine sulfate cannot follow variations of the field at frequencies above 1 Mc,

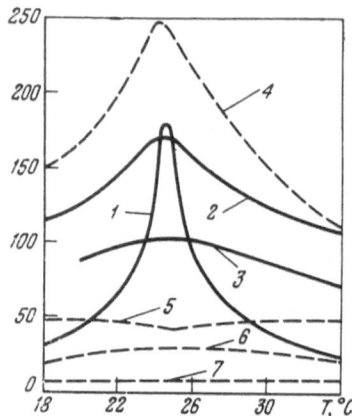

Fig. 185. Temperature dependence of the initial permittivity of Rochelle salt at various frequencies: 1) 10^3 cps (values of ε reduced by a factor of 10); 2) 2×10^8 cps; 3) 3×10^9 cps; 4) permittivity of a "clamped" crystal at frequencies of 1.6×10^5 and 2×10^7 cps; 5) 9.39×10^9 cps; 6) 10^{10} cps; 7) 2×10^{10} and 2.5×10^{10} cps [14b].

and the contribution of the domain polarization to ε (which is very important at low frequencies and represents about 25-50) disappears altogether. On this basis, we may conclude that the relaxation of ε of triglycine sulfate, observed in the 10^5-10^7 cps frequency range, may be associated with domain processes with even greater justification than the dispersion observed for $BaTiO_3$ in the centimeter range of frequencies.

The polarization induced in triglycine sulfate above the dispersion frequencies exhibits relaxation a little above 2.4×10^{10} cps at room temperature, at about 10^9 cps at the Curie point, and considerably above 2.4×10^{10} cps at temperatures over 70°C.

The permittivity of potassium dihydrogen phosphate (KH_2PO_4) obeys the Curie−Weiss law right up to 10^7 cps. Apart from the piezoelectric dispersion of the permittivity, this compound does not exhibit any relaxation of ε_c right up to 10^{10} cps. The Curie temperature, deduced from measurements of ε_c carried out on a clamped crystal at 10^7 cps, gives a value of T_C which is 4° lower than T_C of a free crystal. Since the Curie constants of the clamped and free crystals are the same at high and low frequencies, it follows that the difference between ε_c of free and clamped crystals is independent of temperature.

Measurements on Rochelle salt crystals, carried out above the piezoelectric resonance frequencies, show that the permittivity of this compound obeys the Curie−Weiss law. In this crystal, as in the case of KH_2PO_4, the difference between ε_a of free and clamped crystals is independent of the temperature. The permittivity ε_a of Rochelle salt exhibits dispersion in the frequency range 3×10^8-3×10^9 cps. Thus, according to one report, ε_a at 20°C is 125, 92, and 75 at frequencies of 2×10^8, 10^9, and 3×10^9 cps, respectively. Information on the temperature dependence of ε_a of Rochelle salt near the upper Curie point is given in Fig. 185. It is evident from this figure that at frequencies of about 2×10^{10} cps or higher the value of ε_a is independent of temperature and has a low value (~ 7).

The fall of ε_a, observed in the microwave dispersion region, in the presence of a static electric field and mechanical stresses, indicates that this dispersion is associated with domain processes in Rochelle salt, i.e., with gradual exclusion of these processes from the polarization at increasing frequencies of the applied field. The low value of ε_a at frequencies of $\sim 2 \times 10$ cps indicates also that the principal dispersion region of Rochelle salt (associated with natural vibrations of ions or with lattice dynamics) is close to this frequency.

The low-frequency permittivity of guanidine aluminum sulfate crystals ($\varepsilon_c = 6.1$) is practically independent of the frequency up to 3×10^9 cps.

§6. Dielectric Hysteresis and Changes in Domain Structure

The nonlinear dependence of the polarization on the field (Fig. 173), typical of ferroelectrics, gives rise to a dielectric hysteresis in alternating electric fields (Fig. 112), i.e., the polarization P and the electric field E differ in phase. The dielectric

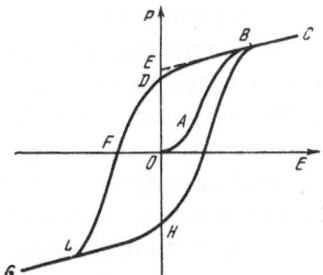

Fig. 186. Dielectric hysteresis loop.

hysteresis loop (Fig. 186) is one of the most important characteristics of a ferroelectric and gives information on its dynamic
polarizability. For this reason, the dielectric-hysteresis method
is widely used to investigate the polarization of ferroelectrics in
strong fields.

An analysis of the hysteresis loop gives clear information
on the role of domain processes in the polarization of ferroelectrics. The loop can be represented schematically as follows. In
very weak fields (OA part of the curve in Fig. 186) the polarization is approximately proportional to the field and is not accompanied by changes in the domain structure. Using the results reported in §4 of the present chapter, we may conclude that, in
fields corresponding to the linear relationship between P and E,
there is practically no growth of new domain nuclei or any motion
of such nuclei at a given field frequency (which represents the duration of application of the field). In the region AB the electric
field causes an effective increase in the polarization. At the point
B the polarization associated with the domains reaches saturation
and the crystal transforms to the single-domain state. In the region BC we again have a linear relationship between P and E,
which represents induced polarization which is not associated with
changes in the domain structure.

The reduction in the value of the field from its maximum at
the point C retraces the original curve down to the point B, but
beyond this point the initial polarization process is not repeated.
When the field is reduced to zero the polarization does not vanish
but becomes equal to the value corresponding to the saturation field (point D on the hysteresis loop). This means that
the single-domain state is almost entirely retained in the absence
of a field and only a few of the oriented domains return to their
initial state. The application of a weak field in the opposite direction lowers the polarization somewhat, which indicates that the
single-domain state is still retained to a considerable degree and
that domain processes are still unimportant. However, when the
field approaches the coercive field OF the polarization decreases
rapidly, passes through zero and, having changed its sign, it again
becomes large. This means that in fields close to the coercive
value the polarization reversal processes are proceeding rapidly:
in these processes the polarization changes its direction. Further
increase of the field, beyond the saturation point, again gives rise

to a linear relationship between P and E, where once more domain processes are inactive. Lowering of the field from the value corresponding to the point L repeats the process represented by the part of the curve BD; the hysteresis cycle ends, via the path LHB, at the point B. During the next period of the field the polarization follows the cycle right round.

The hysteretic nature of the dependence of P on E can be understood on the basis of thermodynamic relationships given in Chap. VI. The relationship between the thermodynamic quantities, given by Eq. (VI.23), is shown graphically in Fig. 187 in the form of the curve ABCD. The region BC of this curve represents an unstable state and, therefore, during polarization reversal, a crystal switches directly from B to E and from C to F. A quantity equal to GH/2 represents the coercive field E_c. However, we must point out that the value of E_c calculated from the thermodynamic relationships is always considerably higher than that observed experimentally. This is because the thermodynamic theory presumes that the whole crystal switches to the opposite polarization as one unit while actually this process involves the creation and growth of new domains (§4 in the present chapter).

Since the process of polarization reversal depends on the field intensity, the value of E_c also depends on the field intensity and its frequency. The general frequency dependence of E_c should be an increase of the coercive field with increasing frequency (at a fixed value of the alternating field amplitude). The value of E_c is increased also by an increase of the field intensity because in stronger fields a greater degree of polarization reversal can be achieved in each half-period and this impedes the nucleation of new domains during the next half-period.

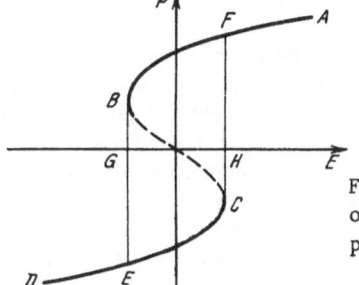

Fig. 187. Polarization as a function of the field for ferroelectrics with a phase transition of the second kind.

The coercive field of a ferroelectric depends on the temperature. A general feature of this dependence is that the coercive
field increases, because of a decrease in the domain wall mobility,
when the temperature is reduced from the Curie point.

The coercive field depends in a complex manner on the quality of a crystal, the nature of its electrodes, and the thickness of
the sample. This dependence is due to the presence of various
surface layers which affect the nucleation and growth of new domains. In general, the coercive field increases when the thickness of a sample is reduced. This point will later be discussed in
more detail.

A. Barium Titanate. The hysteresis loops observed
for BaTiO$_3$ single crystals exhibit definite saturation of the polarization due to domain reorientation: beginning from fields somewhat higher than E$_c$ an increase of the alternating field produces
only a linear (proportional to the field) increase in the polarization.
The hysteresis loops of barium titanate are nearly rectilinear.
The value of E$_c$ lies within the range 500–2000 V/cm.

The relationship between the polarization P of BaTiO$_3$ and
the alternating field intensity E (this relationship governs the shape
of the hysteresis loop and the value of the coercive field) can be
found by assuming that the rate of polarization reversal [which
obeys Eq. (VII.62)] in weak fields depends exponentially on the field
intensity, i.e.,

$$\frac{dP}{dt} = F(P)\, e^{-\alpha/E(t)}, \qquad\qquad (VII.79)$$

where t is the time; α is the temperature-dependent activation

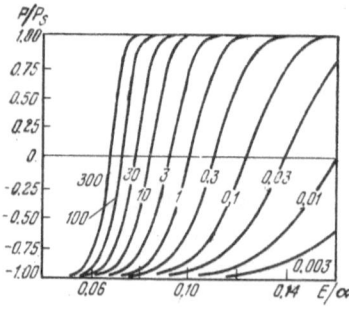

Fig. 188. Dependence of the polarization of
BaTiO$_3$ on the frequency and amplitude of the
applied field. The abscissa represents the
quantity E/α, which varies linearly with time.
The numbers alongside the curves give the approximate values of the rate of rise of the applied
field in relative units. The unit is such a rate
of rise of the field that 1000 V / cm is reached
in 1 msec [53].

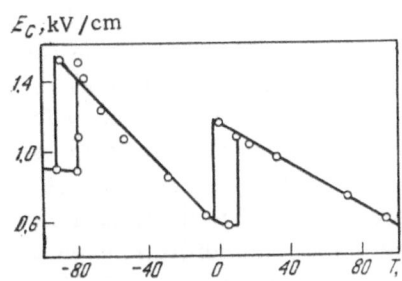

Fig. 189. Temperature dependence of the coercive field of BaTiO₃ [74].

field; F(P) is a function which allows for the fact that the rate of polarization reversal depends on the fraction of the crystal volume which is already reversed. Since $E = E_0 \sin \omega t$, it follows that when $E_0 \gg E_c$ we have

$$E = E_0 \omega t = E't,$$

where E' is the derivative of the field with respect to time. Integration of Eq. (VII.79) gives a dependence shown in Fig. 188. It is evident from this figure that when the external field increases slowly (low frequencies) a crystal can be reversed well before the maximum value of the field E_0 is reached. When the field increases rapidly (high frequencies) the duration of application of the field (the half-period) is insufficient for the completion of the polarization reversal processes and, therefore, the coercive field increases and the loop becomes more rounded. Pulvari and Kuebler [64] have demonstrated experimentally that the dependence $1/E = f (\ln \omega)$ for BaTiO₃ in the case $E_0 = $ const is a straight line for a range of values of ω extending over four orders of magnitude. The same result is obtained also for the dependence $1/E_c = f (\ln E_0)$, corresponding to $\omega = $ const. Because of the dependence of E_c on the field amplitude E_0 and frequency ω, the coercive field cannot be used as a constant characterizing the quality of a given crystal. The experimentally determined temperature dependence of the coercive field of BaTiO₃ is shown in Fig. 189.

There have been several attempts to estimate theoretically the value of the coercive field. Thus, Janovec [48] regards polarization reversal in BaTiO₃ as a process governed by the nucleation and growth of antiparallel domains. According to Janovec, the coercive field is the field corresponding to the onset of an exceptionally rapid rise of the probability of the appearance of anti-

parallel nuclei. He assumes that each such nucleus is in the form
of an elongated half-ellipsoid of revolution. An analysis of the
stability of such a domain nucleus shows that, at certain critical
values of the radius r and length l, the energy U necessary for
the formation of this nucleus has a minimum and that this energy
decreases with increasing external field E. The coercive field E_c
is taken to be that field at which the energy of formation of a critical nucleus is equal to zero. Under these conditions, even thermal fluctuations can generate antiparallel domain nuclei. The appearance of these nuclei shifts a crystal to a metastable state (Fig.
187) in fields weaker than those necessary for the simultaneous
polarization reversal in the whole crystal. The value of E_c obtained by Janovec is 650 V/cm (for an infinitely thick crystal),
which is in good agreement with the experimentally obtained values.

In the present chapter we have already compared the processes of static and pulsed polarization of crystals, accompanied
by the nucleation of new domains, their subsequent motion, etc.
The relationships governing these processes apply also to the polarization in high-intensity alternating fields accompanied by dielectric hysteresis.

Since the structure of the 180° and 90° domain walls in barium titanate is different, the motion of these walls in alternating
fields is also different. As already mentioned, the dipole coupling
along directions perpendicular to the spontaneous polarization axis
is relatively weak compared with the strong coupling along P_s .
This produces unfavorable conditions for the polarization reversal
of a crystal by sideways motion of the 180° walls. Moreover, the
activation energy necessary for the sideways motion of a domain
wall in BaTiO$_3$ is comparable with the total wall energy ($\sim$ 7
ergs/cm^2) since the thickness of this wall is of the order of one
unit cell. Moreover, the wall energy exceeds considerably the
quantity kT. Thus, under these conditions, the sideways motion
of domain walls, caused by thermal activation, is extremely un-
likely. Moreover, an external field which may be even close to
the breakdown value does not play an important role because the
energy EP$_s$, corresponding to the shift of a wall by a distance equal
to one unit cell, represents a small fraction of the wall activation energy. The formation of nuclei of antiparallel domains requires a much lower energy (compared with the activation energy

of a 180° wall) and, consequently, $BaTiO_3$ samples with the 180° walls are switched by the nucleation of new domains and their subsequent growth.

The rate of nucleation of domains in $BaTiO_3$ and in Rochelle salt follows the law (VII.59)

$$\frac{\partial n}{\partial t} \propto e^{-\alpha/E}.$$

According to the observations of Little [54], domain nuclei are in the form of thin wedges. A field $E_0 = 7.5$ kV/cm produces several domains of new polarity in a time interval of the order of 10 μsec, but these domains are rapidly expelled from a crystal when the field is removed.

When the field intensity is increased, the rate and degree of polarization reversal of a sample increase. Wedge-shaped domains of a disappearing (shrinking) component of a twin are frequently observed and these domains are not expelled even by very strong fields. Nuclei of a new component are formed in subsequent cycles at approximately the same points. Evidently this is because domains are nucleated preferentially at crystal defects.

These preferentially formed domains grow along the field direction at a velocity proportional to the field [Eq. (VII.60) and (VII.63)]:

$$v = \mu (E - E_{cr});$$

here, the mobility μ and the critical field E_{cr} are (at room temperature) 2.5 $cm^2 \cdot V^{-1} \cdot sec^{-1}$ and 4000 V/cm, respectively.

These values show that the velocity of the forward motion of domain nuclei in $BaTiO_3$ is considerably lower than the velocity of sound even in very strong fields (higher than 10^4 V/cm).

It follows from the geometry of the 180° domains in $BaTiO_3$ (Chap. IV) that the configuration of these domains is not affected by the external shape of the crystal and, therefore, these domains are not affected by external forces. Consequently, no elastic or piezoelectric hysteresis of the 180° domain walls is observed.

Many investigators have tackled the problem of the motion of the 90° domain walls in the tetragonal modification of $BaTiO_3$

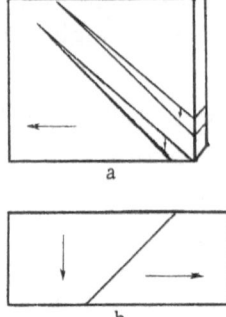

Fig. 190. Configuration
of the 90° domains in
$BaTiO_3$: a) nonequivalent
domains; b) equivalent
domains.

subjected to an external field and the role of the 90° processes in
the polarization reversal. Little [54] demonstrates that the application of an electric field perpendicularly to the spontaneous polarization axis produces 90° wedge-shaped domains at the surface
of a barium titanate crystal (Fig. 190). The critical field for the
nucleation of these domains is again a few kV/cm. In contrast to the 180° domains, the nucleation of the 90° domains
becomes more and more difficult as the number of wedge-
shaped domains increases. This may be due to the electromechanical interaction of the wedges or due to the reduction in the number of points at which new nuclei can be
formed. The maximum velocity of these wedge-shaped domains
approaches asymptotically the velocity of sound. The wedges
begin to grow only at the cathode, which is probably due to the
presence of surface layers (§ 7 of the present chapter). When the
frequency of the sinusoidal field is increased, the amplitude necessary for a given displacement of a domain wall increases as well.
An increase of the frequency of a field of constant amplitude reduces domain wall displacements until oscillation of the walls
stops altogether at frequencies ~ 3 Mc.

Fousek and Brežina [42] have investigated the nature and
motion of a 90° domain wall in an alternating field of 50 cps frequency using a $BaTiO_3$ crystal containing only one such wall (Fig.
190b). They have found that, at voltages V_0 greater than a certain
critical value V_{cr}, the dependence of the domain wall position on
the field intensity produces a hysteresis loop asymmetrical with
respect to the origin of the coordinates. The lowest value of the
field which displaces a 90° wall is 300 V/cm. Oscillations of the

wall increase with increasing value of the voltage. Under certain conditions the equilibrium position of the wall can shift and the wall itself can form a kink.

Fousek and Brežina have interpreted their results from the phenomenological point of view. They have taken into account the inhomogeneity of the field, due to different values of ε_c and ε_a, the dependence of the field on the position of the wall, the difference between the energies of the two domains in a crystal, and the role of the 180° processes. An analysis shows that the nature and motion of a 90° wall are closely associated with the 180° polarization processes. One of the problems is to distinguish these two processes. Direct optical observations of the amplitude of the oscillations of a 90° wall make it possible to estimate the contribution of the 90° processes to the value of ε of a crystal. According to Fousek and Brezina, the 90° processes can increase the value of ε by several hundreds, which corresponds to the value of ε in fields weaker than the coercive field.

The 90° wedge-shaped domains produce considerable internal stresses. Consequently, when the applied field is removed the 90° wedges disappear rapidly.

We note that a 90° wall in $BaTiO_3$ is quite thick (Chap. IV) and, therefore, its displacement by a field requires an activation energy which is only a small fraction of the total energy of the wall. The displacement of the 90° walls by the field may alter the shape of a crystal. Consequently, at field frequencies exceeding the mechanical resonance frequency the domain wall oscillations are in antiphase with the vibrations of the crystal, and they decrease in amplitude. The 90° domains may nucleate and grow under the action of mechanical stresses. However, a centrosymmetric effect, such as a mechanical stress, cannot polarize an unpolarized crystal.

B. Triglycine Sulfate. The hysteresis loops obtained for triglycine sulfate crystals are very nearly perfect rectangles: in a narrow range of fields close to the coercive value almost all the domains reverse their polarization and the induced polarization is very small compared with the spontaneous component. In a 50 cps field of 1500 V/cm amplitude the coercive field E_c is about 400 V/cm (at room temperature). The value of E_c

decreases on approach to the Curie point and at about $-10°C$ it begins to rise rapidly because of a fall in the domain mobility.

A detailed investigation of the dielectric hysteresis of triglycine sulfate crystals has been carried out in the frequency range
0.01-50 cps by Gurevich and Zheludev [4]. The results of their
investigations are summarized in Fig. 191. The nature of the
frequency dependence of E_c (rise of E_c and more rounded hysteresis loops at higher frequencies) for triglycine sulfate is the same
as for $BaTiO_3$ and does not require further comment. The dependence of E_c on the amplitude of the alternating field (E_c increases
with E_0) is particularly strong for triglycine sulfate. This dependence can be understood by recalling that an increase of E_0 increases the degree to which a crystal approaches the single-domain state by providing favorable conditions for the complete ex-
pulsion of domains directed opposite to the field. The nucleation
of new domains in a crystal which consists practically of a single
domain is difficult and this increases the value of E_c. We must
point out also that an increase of E_0 makes the hysteresis loops
approach more closely the rectangular shape. This is because
high-intensity fields of a given frequency provide more favorable
conditions for the effective nucleation of new domains.

Because of difficulties encountered in observations of the
domain structure of triglycine sulfate under dynamic conditions
there is as yet no published information on changes in the domain
structure corresponding to a dielectric hysteresis loop. However,
it can be shown that unipolar blocks of triglycine sulfate crystals
(Fig. 74) have unipolar (displaced) hysteresis loops.

C. Rochelle Salt. Dielectric hysteresis loops can be
observed quite easily between the upper and lower Curie points of
Rochelle salt. The loops approach most closely a rectangle in the
temperature range from +5 to +15°C. The coercive field in this
range of temperatures is about 200 V/cm at a frequency of 50 cps.
The coercive field decreases on approach to either of the Curie
points and the hysteresis loop degenerates to a straight line at
these points. When the frequency of the external field is increased
but other conditions are kept fixed, saturation in the hysteresis
loop requires higher field intensities. At sufficiently high field intensities (ensuring a high rate of domain nucleation and a high wall
velocity) hysteresis loops can be observed at field frequencies up

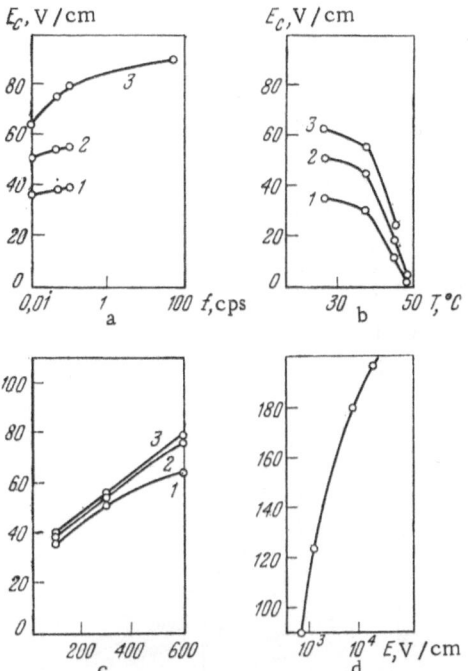

Fig. 191. Dependence of the coercive field E_c of triglycine sulfate on the field frequency (a), temperature (b), and field intensity (c and d). a, b) E = 100, 300, and 600 V/cm for curves 1, 2, and 3, respectively, T = 27.4°C, f = 0.01 cps; c) f = 0.01, 0.05, and 0.1 cps for curves 1, 2, and 3, respectively, T = 27.4°C; d) f = 50 cps, T = 29.5°C [4].

to 10^5 cps. As usual, an increase of the alternating field amplitude of fixed frequency increases the coercive field. It is interesting to note that hysteresis loops (even unsaturated loops) are not observed at low field intensities (a few tens of V/cm) of 50 cps frequency because of strong dispersion processes in Rochelle salt (§4 in the present chapter); under such conditions the field dependence of the polarization is nearly linear.

The relative ease of observation of domains in Rochelle salt makes it possible to associate the various stages of dielectric hysteresis loops with changes in the domain structure. Observations

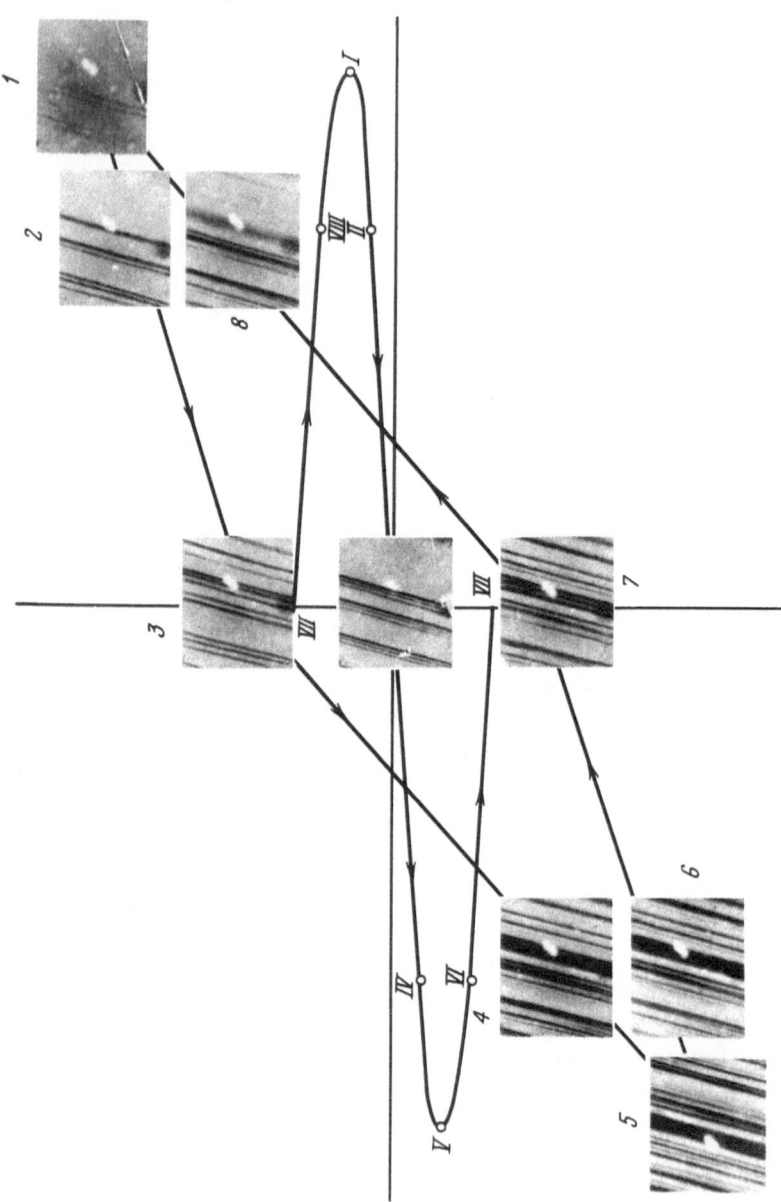

Fig. 192. Changes in the domain structure of Rochelle salt in an electric field of 32 cps frequency [9]. The field amplitude is 36 V/cm. The Roman numbers on the curve representing the sinusoidal dependence of the field on time correspond to the Arabic numbers of the photographs of the domain structure.

carried out at a frequency of 32 cps, using stroboscopic illumination of a crystal, are illustrated in Fig. 192. This figure shows
clearly the domain structure hysteresis and the lag of the polarization reversal process behind the rising field. The difference
between the domain structure at the two extreme states of polarization (which do not quite represent two single-domain states of
opposite orientation) can be seen quite clearly. Analysis of results similar to those shown in Fig. 192, but obtained at a higher
frequency (1000 cps), shows that at this frequency a higher field
intensity is required to obtain the same degree of polarization as
at lower frequencies. On the other hand, at higher field intensities the process of polarization reversal is accompanied not only
by displacements of the walls of existing domains but also by the
effective nucleation of new domains.

The reorientation of domains in Rochelle salt involves a
change in the domain structure and in the shape of the crystal.
Consequently, the dielectric hysteresis of Rochelle salt is affected
strongly by its mechanical state. Moreover, the application of
particular mechanical stresses to Rochelle salt may give rise,
because of the piezoelectric effect, to polarization reversal equivalent to that produced by an electric field (Chap. IX).

D. Other Crystals. A low-frequency coercive field
of KH_2PO_4 has a special feature: during cooling it begins to rise
very rapidly near 60°K (its value at this temperature is close to
2000 V/cm), but the state of a crystal shows no other dielectric
anomalies. This rise of E_c is evidently due to a rapid fall of the
domain mobility at temperatures below 60°K. The domain polarization reversal in KH_2PO_4 can be induced in the same way as in Rochelle salt, i.e., by the application of mechanical stresses which
produce the piezoelectric effect.

The phenomenon of a rise in the coercive field during cooling, which prevents the observation of dielectric hysteresis, is a
feature of several other crystals. Thus, for example, hysteresis
loops cannot be observed for $NaNO_2$ below 100°C (the Curie temperature of this crystal is $T_c \approx 160°C$). At room temperature the
power-frequency coercive field of guanidine aluminum sulfate is
about 1500 V/cm, and it rises very rapidly at temperatures of the
order of −50°C. At lower temperatures the reorientation of domains in guanidine aluminum sulfate can be produced only by the

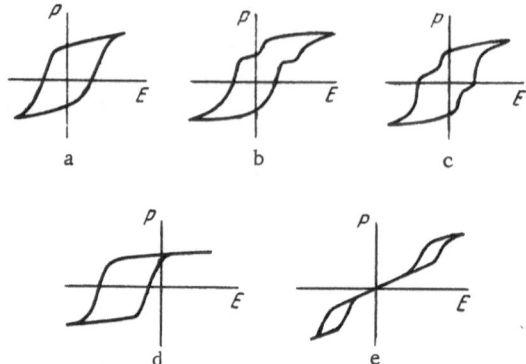

Fig. 193. Distortion of the dielectric hysteresis loop:
a) normal loop; b) ferroelectric consisting of noniden-
tical weakly unipolar blocks; c) ferroelectric consisting of two identical weakly unipolar blocks; d) crystal consisting of a single unipolar block; e) crystal
consisting of two identical strongly unipolar blocks.

application of a static electric field for a long time. Concluding
our discussion of dielectric hysteresis, we must make some comments on distortions of the hysteresis loops, Barkhausen clicks
(jumps), and polarization reversal (switching) noise in ferroelectric crystals.

The most frequently encountered distortions of the hysteresis loops are presented in Fig. 193. Figure 193d shows a displaced hysteresis loop. A distortion of this type can be obtained
by applying a static bias field to a crystal exhibiting a normal hysteresis loop (Fig. 193a). In real crystals a hysteresis loop may
be displaced in samples having natural unipolar blocks (blocks of
domains of the same polarity). The loops shown in Figs. 193b,
193c, and 193e are obtained in those cases when the investigated
sample consists of unipolar blocks of opposite orientations. In
particular, Fig. 193c corresponds to the case in which a sample
consists of two parts of identical area and a weak unipolarity, and
Fig. 193e corresponds to a similar case with strong unipolarity.
Such loops are usually called double loops.

Investigations of the dependence of the polarization or the
polarization current on the field, carried out using sensitive apparatus, show that an increase in the polarization (current) is fre-

quently discontinuous. Such a dependence suggests that the polarization reversal occurs suddenly in small parts of a crystal. Such a region is known as a Barkhausen region (by analogy with the magnetization of ferromagnets) and the polarization reversal in such a region is called a Barkhausen jump.

A Barkhausen jump in a ferroelectric represents an initial rapid motion of a domain across a sample. Thus, the volume of a Barkhausen region should be less than the volume of a domain. In the case of $BaTiO_3$, the volume of a Barkhausen region is estimated to be 10^{-11} cm^3 and in triglycine sulfate it is 10^{-7} cm^3. The existence of Barkhausen regions, their rapid polarization reversal, followed by subsequent "immobility" can be explained by the presence of various defects (growth defects, dislocations, and impurities) which impede the motion of a domain wall. The sudden jumps of domain walls in Rochelle salt have actually been observed experimentally.

Some details of the mechanism of polarization reversal can be deduced from investigations of the noise observed in cyclic polarization reversal (switching) in ferroelectrics. Under these conditions noise can be generated only if the processes taking place are not exactly reproducible from one period to another. This can happen, for example, because of small fluctuations of the moments of single domains during polarization reversal. Under these conditions the periodicity is not perfect and the spectrum contains, in addition to discrete lines, which are harmonics of the fundamental field frequency, an additional continuous component which is known as noise.

The spectral density (G) of fluctuations of the current, considered as a function of the frequency f in regions far removed from the harmonics of the fundamental frequency, can be regarded as a characteristic of the polarization reversal noise. Examples of the noise spectra of $BaTiO_3$ and triglycine sulfate crystals are shown in Fig. 194. It is evident from this figure that the noise spectrum of a ferroelectric can be represented by a frequency f_m, corresponding to the noise maximum. Investigations show that when the polarization reversal frequency is increased the value of f_m increases. This increase of the noise should be compared with the changes in the form of the hysteresis loop observed when the field frequency is increased, and they indicate that the polariza-

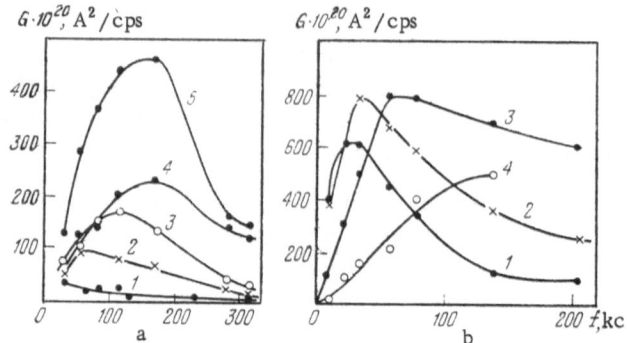

Fig. 194. Spectrum of noise in BaTiO₃ (a) and triglycine sulfate (b) crystals [1]. a) f = 4 kc, $E_{\sim}$ = 1.8 kV/cm, temperature of a crystal: 1) 22°C; 2) 40°C; 3) 70°C; 4) 112°C, 5) 108°C. b) f = 4 kc, $E_{\sim}$ = 0.8 kV/cm, temperature of a crystal: 1) 28°C; 2) 35°C; 3) 43°C; 4) 45°C.

tion reversal process is governed not only by the field amplitude but also by the duration of its application (i.e., by its period). When the amplitude of the polarization reversal (switching) field is increased, the noise spectrum may shift in the direction of higher frequencies and the noise amplitude may increase. This can be understood because an increase of the field produces a closer approach to the single-domain state of a sample and, consequently, a new direction of polarization can be established primarily by the appearance of new domain nuclei which are weakly correlated with the positions of the nuclei generated in the preceding cycles. When the temperature is increased the noise spectra of triglycine sulfate and BaTiO₃ crystals shift in the direction of higher frequencies. Above the Curie temperature the noise amplitude decreases, but there are no special noise singularities at the Curie point. The maximum of the noise spectrum of Rochelle salt corresponds to −9°C. The average dimensions of the regions in which polarization is reversed (by the formation of a single nucleus) can be estimated from the noise spectra, and they amount to 5×10^{-14} and 6×10^{-11} cm³ for BaTiO₃ and triglycine sulfate, respectively. An analysis of the noise spectra shows that when the amplitude of the polarization reversal field and the temperature are increased (i.e., when the Curie temperature is approached) the volume of the polarization reversal regions increases not because of an increase in the number of nuclei but because of enlargement of the regions whose polarization is reversed by the formation of single nuclei.

The contribution of nonperiodic processes to the polarization reversal in single crystals is small compared with the contribution of periodic events. This conclusion follows from changes in the domain structure during polarization reversal observed using stroboscopic illumination. It is evident that the stroboscopic method can give results only if the domain polarization reversal is a periodically repeated process.

§7. Dielectric Nonlinearity

The nonlinear relationship between the polarization and the field (Fig. 173) is one of the main and dominant characteristics of a ferroelectric. In the final analysis, the nonlinearity is due to the domain structure. The clearest manifestation of the nonlinearity is the dielectric hysteresis considered in the preceding section. For this reason ferroelectrics (and antiferroelectrics) are frequently called simply nonlinear dielectrics. However, it does not follow that only ferroelectrics exhibit a nonlinear dependence of the polarization on the field. Nonlinear dependences of P on E are exhibited also by many substances with the relaxational type of polarization (§3 in the present chapter) and these dependences may appear in relatively weak fields. In strong fields even classical linear dielectrics (which exhibit only the elastic displacement types of polarization) can also have a nonlinear relationship between P and E. A similar nonlinearity may be also found in single-domain ferroelectrics; such nonlinearity is associated not with the spontaneous polarization but with the elastic polarization and it is independent of domains or their reorientation. We shall now consider in detail the effects of dielectric nonlinearity observed in ferroelectrics.

A. Basic Thermodynamic Relationships. In the case of a single-domain ferroelectric, it is simplest to consider the relationship between the spontaneous polarization and the field defined by Eq. (VI.23):

$$E = \frac{\partial G}{\partial P_s} = \alpha^* P_s + \beta P_s^3 + \gamma P_s^5 + \cdots$$

If we retain only two terms on the right-hand side of Eq. (VI.23), the nonlinearity is represented by the coefficient β. When $\beta = 0$ the ferroelectric is linear because, in this case, the relationship between P and E (at a given temperature and pressure) is linear. The permittivity of a single-domain ferroelectric measured along the direction of polarization, ε_E , is essentially reversible, i.e., it is the permittivity which is measured using weak alternating fields during the simultaneous application of a constant (bias) field $E_=$. Thus, the permittivity along the spontaneous polarization direction of a single-domain crystal (the reversible permittivity) is found from the relationship

$$\varepsilon_E = 1 + 4\pi \frac{dP}{dE_\sim}. \qquad\qquad (VII.80)$$

Using the relationship between P_s and E given by Eq. (VI.23) and assuming that $\varepsilon \gg 1$, we find that Eq. (VII.80) yields

$$\beta = \left(\frac{4\pi}{\varepsilon_E} - \alpha^\bullet\right) \frac{1}{3P_s^2}. \qquad\qquad (VII.81)$$

In a multidomain ferroelectric the permittivity measured in weak alternating fields differs from the reversible value. When the constant bias field is increased during the measurements, a crystal is gradually transformed to the single-domain state, domains cease to be active in the polarization-reversal processes, and the reversible permittivity decreases.

In order to estimate the nonlinearity coefficient β of a multidomain ferroelectric we can use Eq. (VII.81), in which P_s must be replaced by the total polarization measured in a given experiment. The value of the coefficient β for multidomain ferroelectrics can be found also by measuring not the reversible but the ordinary permittivity ε and the polarization P. In this case, the expression

$$\varepsilon = 1 + 4\pi \frac{P}{E} \qquad\qquad (VII.82)$$

and Eq. (VI.23), combined with the condition $\varepsilon \gg 1$, yield the relationship

$$\beta = \left(\frac{4\pi}{\varepsilon} - \alpha^\bullet\right) \frac{1}{P^2}. \qquad\qquad (VII.83)$$

The coefficient β can be found also by investigating the nonlinear properties of ferroelectrics above the Curie temperature, i.e., in the nonpolar modification. Retaining only two terms in the expansion of Eq. (VI.23), we obtain

$$\frac{dE}{dP} \cong \frac{4\pi}{\varepsilon} = \alpha^* + 3\beta P_s^2.$$ (VII.84)

Since above the Curie temperature $P_s = 0$, it follows that $\alpha^* = 4\pi / \varepsilon_0$ (where ε_0 is the permittivity near the transition temperature but on the nonpolar side of this temperature). Substituting the value of α^* in (VII.84), we obtain

$$\frac{4\pi}{\varepsilon_P} - \frac{4\pi}{\varepsilon_{P=0}} = 3\beta P^2,$$ (VII.85)

where ε_P is the permittivity of the polar phase. We can easily see that the last relationship is identical with Eq. (VII.81) if we regard the spontaneously polarized state as that in which (because of the presence of the spontaneous polarization) the permittivity ε_P is equivalent to the reversible permittivity ε_E .

Ferroelectrics which undergo a phase transition of the first kind obey, in accordance with Eqs. (VI.29) and (VI.30), the following relationships

$$P_s^2 = -\frac{3}{4}\frac{\beta_0}{\gamma_0},$$ (VII.86)

$$\alpha_0^* \gamma_0 = \frac{3}{16}\beta_0^2,$$ (VII.87)

which can be employed to determine the coefficient β_0 using the values of P_s and α_0^* measured at the transition point. Determinations of the coefficients β for various crystals have been carried out by many authors. There is a fairly good agreement between the values of the coefficients obtained in different experiments. The coefficient β of a free crystal of $BaTiO_3$ decreases linearly with temperature (between 5 and 160°C) and its value is -2.5×10^{-13} cgs esu at the Curie point (120°C) and -7×10^{-13} cgs esu at room temperature. Extrapolation shows that the nonlinearity of $BaTiO_3$ disappears near 175°C. The most reliable value of the coefficient β for triglycine sulfate is 7.7×10^{-10} cgs esu near the phase transition point (49°C). The coefficient β for a free crystal of Rochelle salt is 6×10^{-8} cgs esu. Thus, out of these three ferroelectrics the strongest nonlinearity is exhibited by Rochelle salt, which is followed by triglycine sulfate and $BaTiO_3$.

We have just established [Eqs. (VII.81) and (VII.85)] the relationship between the reversible permittivity and the nonlinearity of a ferroelectric. The greater the change in the permittivity under the action of a bias field, the stronger is the nonlinearity of a crystal. However, it must be pointed out that the application of a constant field to a ferroelectric crystal also shifts its Curie temperature.

The problem of the shift of the Curie points of some ferroelectrics has been discussed partly in Chap. V in connection with the electrocaloric phenomena. We must add here that a constant external field affects differently crystals with phase transitions of the first and second kind. In the case of phase transitions of the second kind, the application of a constant (static) field broadens the phase transition along the temperature scale, and the temperature dependence of the spontaneous polarization P_s (Fig. 195a) is so smooth that we can hardly speak of the Curie point (this effect is due to the induced polarization of the paraelectric phase). In moderately strong fields a phase transition of the first kind does not change its nature but the Curie temperature shifts in the direction of higher temperatures (Fig. 195b). Once again, it is difficult to find the Curie temperature (defined as the temperature at which the nonpolar phase changes to the polar modification) because a crystal is already polarized in the "nonpolar" phase due to the influence of the external field. When the external field intensity is increased, a phase transition of the first kind loses gradually its abrupt nature and broadens over a range of temperatures.

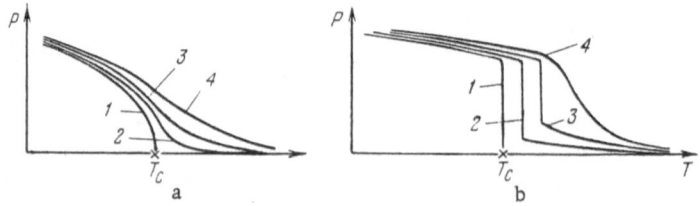

Fig. 195. Temperature dependences of the polarization near phase transitions of the second (a) and first (b) kind in the absence of a bias field (curve 1) and in the presence of such a field (curves 2-4). Curves 2-4 represent fields $E_2 < E_3 < E_4$, respectively.

B. Other Nonlinear Characteristics. Nonlinear properties in ferroelectrics are manifested in many experimentally observed dependences. We have already mentioned the nonlinear relationship between P and E, observed in experiments with quasistatic fields (Fig. 173), as well as the hysteresis of the dependence of P on E in alternating fields. In the preceding subsection we have discussed the reversible permittivity and its dependence on the bias field, which also belongs to the category of nonlinear properties of ferroelectrics.

Apart from the coefficient β of Eq. (VI.23), the nonlinear properties of ferroelectrics are often described by other parameters. In this connection we must stress a point mentioned many times before: the permittivity depends on the field intensity. This can be seen first of all from the dependence of α on E, defined as the derivative dP/dE (Fig. 173). * It is usual to measure either the field dependence of the effective permittivity ε_{eff} or the dependence of the permittivity on the first harmonic of the field, ε_1. The effective permittivity is estimated from the capacitance deduced from the polarization of the capacitor at a given amplitude of the field.

Typical field dependences of ε_{eff} are given in Fig. 196a for some ferroelectric crystals. It is evident from this figure that in weak fields the value of ε_{eff} is comparatively small and that it reaches its maximum values in alternating fields of amplitude close to the coercive field. This is easily understood because the process of polarization reversal is most effective in such fields. An increase of the applied alternating field intensity above the coercive field value does not produce such an effective increase in the polarization and there is a corresponding fall of ε_{eff} (or ε_1). Curves similar to those shown in Fig. 196a are sometimes used to estimate the nonlinearity as a ratio of the maximum value of ε_{eff} and the initial value of ε_{eff} in weak fields.

We can easily see from Fig. 196a that the reversible permittivity ε_E varies most rapidly when it is recorded in an alternating field of intensity corresponding to a maximum of ε_{eff}. The reversible permittivity of some ferroelectrics is shown in Fig. 196b.

---

* We are speaking here of the domain nonlinearity, in contrast to what we can call the "intrinsic" nonlinearity.

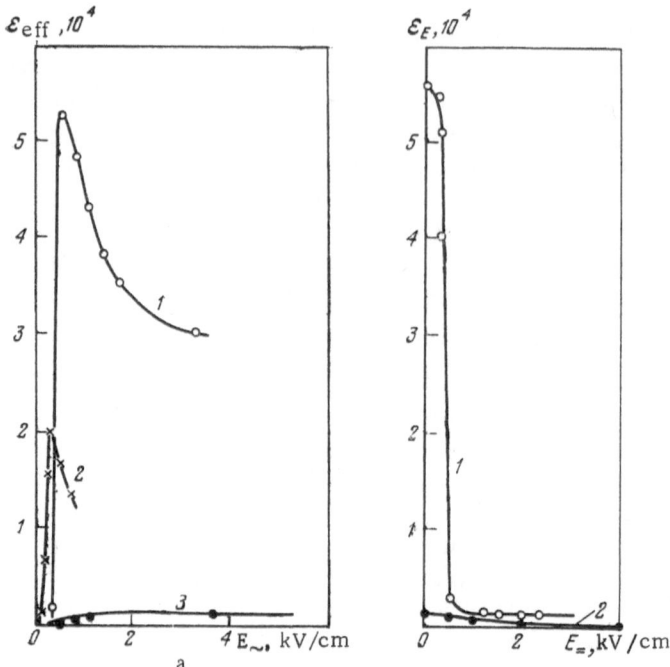

Fig. 196. Nonlinear properties of some ferroelectrics: a)
dependence of ε_{eff} on the intensity of an alternating
field $E_\sim$ used in measurements carried out on $BaTiO_3$ (1),
triglycine sulfate (2), and guanidine aluminum sulfate (3)
crystals; b) dependence $\varepsilon_E(E_=)$ for $E_\sim$, corresponding to
the maxima of ε_{eff} in Fig. 196a for crystals of $BaTiO_3$ (1)
and guanidine aluminum sulfate (2) [7].

We can also use the following measure of the nonlinearity

$$N = \frac{1}{\varepsilon_{eff}} \frac{d\varepsilon_{eff}}{dE_=} . \qquad (VII.88)$$

which represents the relative effectiveness of the suppression, by
an external constant field, of the influence of the domain structure on the polarization processes. N_{max} is about 80 for triglycine sulfate and Rochelle salt, and 30 for $BaTiO_3$.

The nonlinear properties depend on the thickness of a sample because of the influence of surface layers on the polarization
characteristics of a crystal. When the thickness is reduced, a

maximum of ε_{eff} is reached at higher field intensities (because
the coercive fields are higher) and the nonlinearity decreases.

C. Permittivity Measured during Polariza-
tion Reversal. Some ferroelectrics (BaTiO$_3$, triglycine sulfate, lithium trihydrogen selenite, and others) exhibit a strong rise
of the permittivity (in a weak field) when ε is measured during
polarization reversal in a crystal by a strong sinusoidal or pulsed
field. The permittivity measured in this way depends on the frequency. Experimental data indicate an increase of ε right up to
frequencies of 100 Mc. An example of the frequency dependence
of the increment in the permittivity $|\Delta \varepsilon|$ of triglycine sulfate is
shown in Fig. 197. It is evident from this figure that the slope of
the frequency dependence changes sharply at a frequency of 10^7
cps. Below this frequency the increment varies with the frequency
ν as $1/\nu_r$ but above this frequency the increment is proportional
to $1/\nu^2$.

According to Fatuzzo [38], the permittivity ε increases during polarization reversal because of oscillations of the charge
(which is equivalent to the divergence of the polarization) in the
walls of domains growing in an alternating field. The frequency
$\nu = \omega/2\pi = 10^7$ cps, corresponding to a change of the slope in the
frequency dependence, marks the point at which the damping force
on the wall of a domain nucleus, $\gamma \dot{x}$, is equal to the restoring force
$\omega^2 x$. Below 10^7 cps the dominant component of the permittivity of

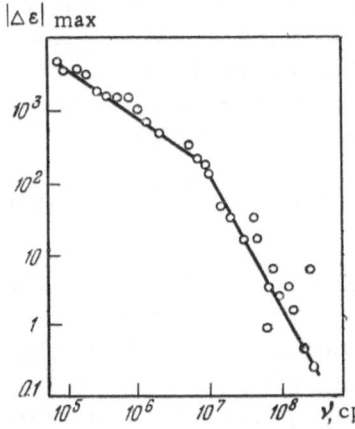

Fig. 197. Dependence of the incre-
ment in the permittivity $|\Delta\varepsilon|$, mea-
sured during polarization reversal in
triglycine sulfate, on the frequency
of the applied alternating field [38].

triglycine sulfate is the imaginary part ε'' but above this frequency the real part ε' predominates. The mass of the domain walls in triglycine sulfate, calculated from the value of the frequency corresponding to the kink in Fig. 197, is 4×10^{-7} g/cm^2, which is three orders of magnitude larger than the corresponding domain wall mass for BaTiO$_3$ calculated theoretically by Kittel [52].

§8. Surface Layers in Ferroelectric Crystals and Their Influence

Experimental observations indicate the existence of surface layers with properties differing from those of the bulk of a crystal. The nature of these layers is not yet fully understood. We are not yet sure whether such layers occur only in ferroelectric crystals.

X-ray investigations, carried out on BaTiO$_3$ crystals by Känzig and his colleagues [24, 51], have established that the structure of surface layers about 100 Å thick differs from the bulk structure. Near the Curie point (below 120°C) the differences between the structure of the surface layers and the bulk of a crystal are smaller, indicating a possible tetragonal structure of surface layers in BaTiO$_3$ above the Curie point. Känzig has suggested that these layers can be regarded as ion-depleted Schottky layers in which impurity concentrations are 10^{18} cm^{-3} and electric fields are about 10^5-10^6 V/cm.

Mertz [59] has established experimentally that the coercive field during polarization reversal in BaTiO$_3$ depends strongly on the thickness of a sample. The polarization reversal time and the coercive field increase with decreasing thickness of a crystal, in a manner similar to an increase of the activation field α of Eq. (VII.62) (Fig. 198). The dependence of the activation field α on the thickness of a sample d can be represented in the form

$$\alpha = \alpha_\infty \left(1 + \frac{d_0}{d} \right), \qquad\qquad (\text{VII.89})$$

where α_∞ is the activation field for an infinitely thick crystal and $d_0 = 10^{-2}$ cm. The dependence represented by Eq. (VII.89) is attributed by Merz to the existence of surface layers of thickness d_s of unknown composition and of much lower permittivity ε_s than the volume (bulk) permittivity ε_v. Taking into account the role of

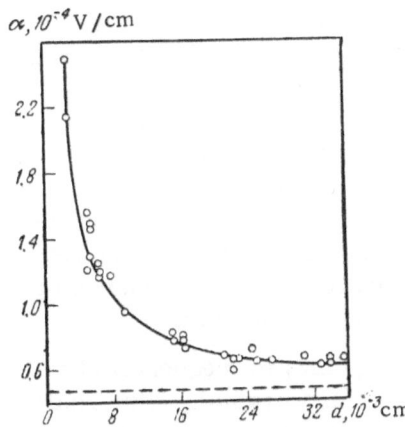

Fig. 198. Dependence of the activation field α on the thickness of a BaTiO$_3$ crystal [59].

the surface layers as dividers of the applied voltage and assuming that ε_s = 5 but ε_v = 150, Merz estimates the thickness of surface layers to be d_s = 10^{-4} cm (these layers are regarded as pure capacitances without losses).

It is mentioned in §4 of the present chapter that, according to Miller and Savage [61a, 61c], the velocity of the sideways motion of domain walls in BaTiO$_3$ is given by Eq. (VII.61):

$$v = v_\infty e^{-\delta/E}.$$

The activation field α and the field δ are similar, at least for metallic electrodes. The similarity of the fields α and δ is manifested by their identical dependences on the thickness of a sample. It has been found that δ, like the activation field of (VII.89), depends on the thickness of a sample d:

$$\delta = \delta_\infty \left(1 + \frac{d_0}{d}\right). \qquad (VII.90)$$

According to estimates of Miller and Savage, d_0 is 5×10^{-3} cm, which is half the value of d_0 obtained by Merz. However, bearing in mind the very large difference between the polarization reversal (switching) times reported by Merz (about 10 μsec) and Miller and Savage (about 1 sec), the agreement between the values of d_0 obtained from these two equations must be regarded as very satisfactory.

The presence of tetragonal spontaneously polarized layers in BaTiO$_3$ above the Curie point (120°C) has stimulated investigations of pyroelectric properties of these crystals. Such properties have indeed been found above the Curie point by Chynoweth [31]. It follows from these observations of the pyroelectric effect that an appreciable polarization may exist above the Curie point in the absence of an external field. Such polarization may be attributed to the presence of a space charge in a crystal. Chynoweth has measured the resistance and estimated that the thickness of space-charge layers near the surface is 3×10^{-5} cm.

The shift of hysteresis loops (Fig. 193) is frequently attributed to the existence of surface layers in a crystal. A more detailed study of this effect shows that such surface layers are only indirectly responsible for the shift of the loops: the shift is observed when the symmetry of the positions of these layers is disturbed. Such a disturbance may be caused, for example, by the use of different electrodes (Fig. 199).

Investigations of the birefringence of BaTiO$_3$, induced by an external field above the Curie temperature, indicate that space-charge layers are formed near the electrode surfaces. These layers reduce the field within a sample. These experiments also show that the thickness and lifetime of the space-charge layers depend on the applied field. According to Triebwasser [73], these layers are ion-depleted Schottky layers with a donor concentration of about 10^{19} cm^{-3}, a permittivity of about 200, and a thickness of 5×10^{-6} cm.

The dependence of the coercive field on the thickness of a sample is explained by Drougard and Landauer [36] in a different way. Drougard and Landauer assume that the displacement current between the electrodes should be continuous during the polarization reversal process. In the case of BaTiO$_3$, whose spontaneous polarization is high (26 μC/cm^2), this condition can be satisfied only when a surface layer of low permittivity switches its polarization like the rest of the sample. A nonswitched layer with a low value of ε (about 5) would require fields of the order of 10^8 V/cm for a change of the polarization by $2 \times 26 = 52$ μC/cm^2, and such fields are higher than the breakdown strength of barium titanate. Hence, Drougard and Landauer conclude that layers about 10^{-4} cm thick (found by Merz and Chynoweth) should undergo the

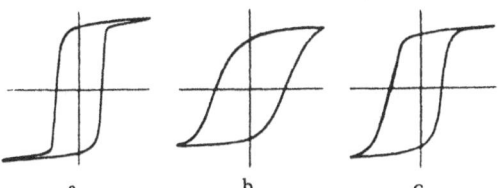

a   b   c

Fig. 199. Hysteresis loops of a $BaTiO_3$ crystal obtained using different electrodes: a) both electrodes consisting of a solution of LiCl; b) both
electrodes made of indium; c) one electrode
made of LiCl solution and the other (on the "right")
made of indium. The crystal was 1.9×10^{-2} cm
thick; the field intensity was 3 kV/cm; the frequency of the field was 50 cps [50].

same polarization reversal process as the rest of a crystal. They
point out that it is difficult to imagine a layer whose polarization
would be similar to that of the bulk of a crystal but whose permittivity would be considerable lower. On this basis, Drougard
and Landauer suggest that the postulate of "thick" ($\sim 10^{-4}$ cm) layers be replaced by an assumption that the surface layers are
much thinner and that they do not suffer polarization reversal on
application of a field. The existence of these very thin layers is
supported by the investigations of Drougard and Landauer, who
have established (by x-ray diffraction) that layers on the surface
of $BaTiO_3$ cannot be thicker than 50 Å.

Drougard and Landauer [36] have considered the influence of
the surface charge density $2P_s$, immediately behind a moving 180°
domain wall at the boundary between a surface layer and the bulk
of a crystal. The surface layer is assumed to be conducting, which
means that the charge may be dispersed or neutralized in some
characteristic time τ. (It is assumed that τ is sufficiently short
and that it ensures neutralization of the charge which has appeared
in the polarization reversal preceding a given cycle.) If a domain
wall moves sufficiently rapidly, it may leave behind an undispersed
charge $2P_s$ (here the subscript "s" denotes the spontaneous polarization). For films whose thickness d_s (here the subscript "s" denotes parameters relating to surface layers) is much smaller than
the thickness of the whole crystal d, Drougard and Landauer obtain

$$E = \frac{V}{d} = E_\mathrm{v} + 8\pi P_s d_s / \varepsilon_s d \qquad \text{(VII.91)}$$

(where V is the applied voltage), which yields the relationship

$$d_0 = 16\pi P_s d_s / \varepsilon_s E. \qquad \text{(VII.92)}$$

Using the values $d_0 = 10^{-2}$ cm, $E = 1000$ V/cm, $\varepsilon_s = 5$, Drougard and Landauer estimate the surface layer thickness d_s to be 4 Å (one unit cell). However, Drougard and Landauer are unable to explain why almost the same dependence of the activation field on the thickness is observed at low and high velocities of domain walls. We would expect that the charge behind a very slowly moving wall would be dispersed completely and this would give rise to a dependence of the activation field on the thickness different from that observed at higher domain wall velocities.

Miller and Savage [61d] assume that $\varepsilon_s = 100$ (rather than 5) and estimate the surface layer thickness d_s to be 80 Å. They show that under these conditions the field in the layer would be fairly high ($\sim 3 \times 10^6$ V/cm) and could give rise to a special type of electroluminescence. This was indeed observed by Harman [45].

Fatuzzo and Merz [40] suggest the following hypothesis. It is assumed that a surface layer in $BaTiO_3$ consists of mechanically stressed and electrically polarized (with shifted hysteresis loops) barium titanate. The stresses and the polarization reduce the permittivity and explain the low value of ε. Since such a layer is part of a $BaTiO_3$ crystal, the requirement of continuity of the electric displacement between the bulk of a crystal and the layer can be satisfied quite easily. According to Fatuzzo and Merz, the electric polarity (manifested by a shift of the hysteresis loop) and the mechanical stresses (piezoelectric effect) are due to Schottky-type layers (this is in agreement with the first suggestions of Känzig). The existence of strong electric fields in the thin layers near the surface should give rise to a higher concentration of the c domains compared with the a domains in $BaTiO_3$. An investigation of x-ray reflection, carried out by Subbarao et al. [70], indicated that the fraction of the c domains at the surface was indeed greater than that which would follow from a random distribution. Moreover, they found that the number of such domains can be altered by various treatments (crushing, etching, or irradiation).

A major contribution to studies of surface layers in BaTiO$_3$ crystals has been made by Czechoslovak physicists. Janovec [49] considers theoretically the conditions for the existence of anti-parallel domains on the surface of a tetragonal single-domain crystal of BaTiO$_3$ in the presence of a surface layer. Janovec uses a very simple model of a surface layer, 10^{-6}–10^{-4} cm thick, with a field within the layer directed toward the interior of the crystal and a potential drop across the layer of about 1 V. Janovec shows that, in that part of the crystal in which the direction of the field in the layer is opposite to the direction of the spontaneous polarization in the bulk of a crystal, conditions are favorable for the existence of the antiparallel (180°) domains. An estimate of the dimensions of such domains gives a diameter of about 10^{-4} cm and a length of 4×10^{-3} cm, in satisfactory agreement with the ex-perimental data.

The results of Janovec can be used successfully to explain various properties of BaTiO$_3$. In particular, etching removes fast-er those regions in which the spontaneous polarization vector is directed outward (positive ends of domains) and the field in the surface layer is directed inward. This corresponds to the presence of a positive space charge in the surface layer. The presence of a surface layer with a field directed normally to the surface fa-vors the experimentally observed preferred antiparallel orienta-tion of domains at the surface of barium titanate. The presence of antiparallel domains in the surface layer and of the inward field in this layer explains why polarization reversal in BaTiO$_3$ crys-tals always begins at the positive electrode.

Dvořák [37] has investigated theoretically whether a surface layer in single crystals could be due to Schottky-type defects. Us-ing a simplified ionic model, Dvořák calculates the potential dif-ference in a surface layer and the concentration of oxygen vacan-cies in the cubic modification of barium titanate as a function of temperature. He finds that the potential difference is independent of temperature in the range 200–600°C and the field is directed in-ward, indicating a positive space charge in the surface layer. Dvořák predicts that at 400°C the surface layer thickness should be 8×10^{-5} cm, which is in agreement with the experimental data. At this temperature the concentration of oxygen defects is 2×10^{14} cm^{-3}. However, Dvořák has been unable to predict correctly the surface layer thickness at lower temperatures. He attributes this discrepancy to a possible influence of impurities and "frozen-in" defects formed during crystal growth at high temperatures.

Coufová and Arend [32-34] have carried out experimental investigations of surface layers formed during the growth and treatment of $BaTiO_3$ crystals under various conditions. They have investigated single crystals prepared by crystallization from the $BaCl_2 - BaO - BaTiO_3$ system and have found that the optical absorption coefficient depends on the sample thickness. This dependence can be explained by the presence of surface layers. Annealing of these crystals in media with different concentrations of oxygen indicates that the surface layers consist of nonstoichiometric oxygen-deficient $BaTiO_3$. Studies of the absorption spectra of oxygen-deficient large crystals indicate the presence of two bands with maxima at 0.48 and 0.64 μ. The second band is attributed by Coufová and Arend to the capture of an electron by an oxygen vacancy.

The presence of surface layers in $BaTiO_3$ crystals, exhibiting a dependence of the optical absorption on the thickness of the layers, is demonstrated by Coufová and Arend by gradual etching away of the surface layers until the absorption ceases to depend on the thickness. For crystals 0.075-0.178 mm thick the thickness of the surface layers is 6×10^{-4} cm. Coufová and Arend have also investigated the dependence of the permittivity of unetched $BaTiO_3$ crystals on their natural thickness. They have found that the value of ε decreases with decreasing thickness, but successive etching away of the surface layers increases ε. Assuming that the permittivity ε of the surface layers differs from ε of the bulk of a crystal, Coufová and Arend have obtained good agreement between the experimental and calculated data. The thickness of the surface layers in these crystals is found to be 11.5 μ and the permittivities ε of the layers and of the bulk crystals at 60°C are about 200 and 1600, respectively.

Glogar and Janovec [43] have investigated the dependence of the coercive field E_c on the thickness d of $BaTiO_3$ crystals. These authors measured the coercive fields of single-domain crystals, whose thickness was reduced by gradual etching away of the surface in H_3PO_4. They used crystals grown from the $BaTiO_3 - KF$ system. A saturated aqueous solution of LiCl was used as the

electrode material. It was found that the initial etching increases
strongly the coercive field ($\partial E_c / \partial d \approx 1.5 \times 10^5$ V/cm^2). After
etching away layers thicker than (1-2) $\times$ 10^{-3} cm, the coercive
field increased much more slowly ($\partial E_c / \partial d \approx 7 \times 10^3$ V/cm^2).
This increase in the coercive field is explained by Glogar and Ja-
novec using Dvořák's calculations [37]. In their opinion, the rise
of the coercive field E_c is associated with the fact that, as the surface layers are etched away, the external field is no longer aug-
mented by the inward field in the surface layers. The removal of
defects present in the surface layers has a similar effect because it tends to produce less favorable conditions for the nucleation of domains during polarization reversal.

A modified view of surface layers in BaTiO$_3$ crystals is given
by Brežina and Janovec [29]. These authors have been able to obtain a better agreement between their calculations and the experimental dependence of the coercive field on the natural thickness of
a crystal by assuming that the permittivity in the surface layer
is not constant but increases from the surface toward the interior
of a crystal. Considering the dependence of the coercive field on
the thickness in the form

$$E_c = E_\infty + \frac{\gamma}{d} \tag{VII.93}$$

(where E_∞ is the coercive field of an infinitely thick crystal, d is
the thickness of the crystal actually considered, and γ is a constant depending on the crystal), Brežina and Janovec estimate the
value of the constant γ to be 1.6 V.

The influence of surface layers on the 180° polarization reversal in BaTiO$_3$ crystals has been investigated by Brežina and
Fotchenkov [28]. These authors have demonstrated that the application of an alternating field to a crystal with surface layers produces a large number of antiparallel domains which exhibit only a
slight sideways motion of the walls. In crystals from which the
surface layers are removed the polarization reversal is due to the
presence of a small number of domains exhibiting mainly the sideways motion. Prolonged application of a field of 10-15 kV/cm acts
in the same way as etching away of both sides of a crystal surface.
It is found that a maximum shift of the 180° domain walls in etched
crystals occurs along the a axis and the minimum shift along a direction making an angle of 45° with the a axis. This produces domains of nearly square shape with concave sides.

Brežina and Fotchenkov consider the mechanism of polarization reversal taking into account the influence of the surface
layers but they do not share the views of Glogar and Janovec on
the role of the electric field in the surface layers. According to
Brezina and Fotchenkov, the importance of the surface layers in
polarization reversal is due to the fact that these layers have a
low (compared with the crystal bulk) permittivity and, therefore,
a high electric field intensity produces antiparallel domains which
act as nuclei during polarization and are responsible for the reduction of the coercive field. Brežina and Fotchenkov have not
observed the unipolarity of the polarization reversal process predicted by Glogar and Janovec: they find that the nature of polarization reversal is independent of whether the field in the surface
layer opposes or enhances the external field.

So far, we have discussed only the surface layers in $BaTiO_3$
single crystals and we have shown that these layers are primarily
due to oxygen defects. Phenomena which may be attributed to surface layers are observed also in other ferroelectrics. Thus, Sonin
and Gladkii [20] have investigated the dependence of the polarization reversal parameters of triglycine sulfate and found that the
coercive field, spontaneous polarization, reversal time, and activation field α all depend on the crystal thickness. In particular,
they have demonstrated that the coercive field increases when the
crystal thickness is reduced by grinding. This dependence is observed from thicknesses of about 6×10^{-2} cm. It is most likely that
at these thicknesses the electrical properties of triglycine sulfate
are affected by the surface layers which are formed during various treatments of these crystals and which have a defective structure. According to Sonin and Gladkii, the resistivity of these layers is high and they are either ferroelectric (with the spontaneous polarization perpendicular to the ferroelectric axis) or nonferroelectric; their origin is attributed to impurities or defects
captured or generated during grinding. Pulvari and Kuebler [64]
have studied triglycine sulfate crystals with well-etched surfaces
and have found that the polarization reversal parameters are independent of the thickness. Wieder [74] reports that the polarization reversal parameters of guanidine aluminum sulfate depend on
the thickness of these crystals.

The problem of surface layers in ferroelectric crystals is
closely associated with the phenomenon of fatigue. The fatigue of

some ferroelectrics is evidently due to space charge. Some (if not all) ferroelectrics exhibit a gradual reduction of the charge switched in polarization reversal after prolonged application of an alternating field causing such reversal; finally, an alternating field fails to reverse the polarization. The question arises whether this fatigue or decay effect is due to fewer and fewer domains being nucleated because of the formation of surface layers or whether it is due to domain wall motion being impeded by the presence of space charge in the crystal interior. There are grounds for assuming that the internal space charge is responsible for the fatigue phenomenon in $BaTiO_3$. It is assumed that large currents flowing through barium titanate crystals during polarization reversal cause an electro-chemical reaction which reduces the charge of Ti^{4+} ions to Ti^{3+}, which increases the space charge. This may explain the blue color acquired by barium titanate crystals under the action of a constant current. Arend and Coufová [25] have demonstrated that $BaTiO_3$ becomes blue not only under the action of a constant field but also after heating in a hydrogen atmosphere. This indicates that the color is due to the lower charge of the titanium ions.

§ 9. Phase Transitions and Polarization of Antiferroelectrics

Changes in the crystal structure, the domain structure, and the thermodynamic theory of the principal properties of antiferro-electrics are discussed in Chaps. II–IV and VI. Here, we shall start with some data on phase transitions in antiferroelectrics and shall then describe their polarization (mainly near phase-transition points).

A. Phase Transitions. The temperature dependence of the specific heat of ammonium dihydrogen phosphate (Fig. 200a) shows that the phase transition from the paraelectric to the anti-ferroelectric modification is accompanied by the absorption of heat, i.e., it is a transition of the first kind. The values of the heats and entropies of transitions of some antiferroelectrics of the $NH_4H_2PO_4$ type are given in Table 41. The high value of the transition entro-py and the fact that it is practically independent of the transition temperature of antiferroelectrics of this type indicate that these transitions are associated with the ordering of structure elements.

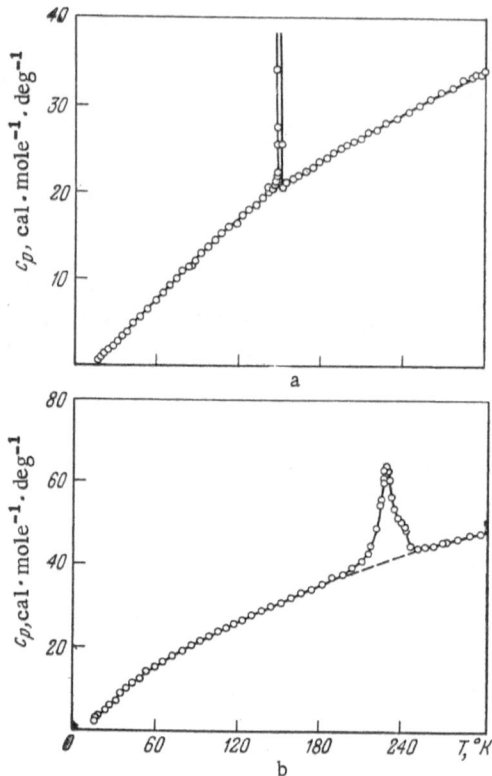

Fig. 200. Temperature dependence of the specific heat at constant pressure for $NH_4H_2PO_4$ (a) [69] and $Ag_2H_3IO_6$ (b) [68].

Figure 200b shows the temperature dependence of the specific heat of a crystal of $Ag_2H_3IO_6$ subjected to a constant pressure.

A phase transition of the first kind in $PbZrO_3$ is also accompanied by a specific-heat anomaly, similar to that shown in Fig. 200a for $NH_4H_2PO_4$. The heat of a phase transition in $NaNbO_3$ at 460°C is small. However, the heats of three transitions in WO_3 are large (Table 41). The absolute values of changes in the entropy of $PbZrO_3$ and WO_3 indicate that their antiferroelectric phase transitions should be of the order-disorder type.

B. Polarization. The published data on the polarization of antiferroelectrics have been obtained mainly in weak and strong fields of low frequencies (tens of cycles per second).

TABLE 41. Heats and Entropies of Transition
for Some Antiferroelectrics [11]

Substance	Transition temperature, °C	Total heat of transition, cal/mole	Entropy of transition, cal·mole⁻¹. deg⁻¹
$NH_4H_2PO_4$	—125	154	1,05
$NH_4H_2AsO_4$	—57	220	1,02
$(NH_4)_2H_3IO_6$	—19	350	1,40
$Ag_2H_3IO_6$	—43	240	1,04
$PbZrO_3$	230	440	0,87
$NaNbO_3$	467	50	0,067
WO_3	~ — 13	*	
	~ 727	450	0,45
	~ 897	280	0,24

*Only qualitative observations were made.

Investigations in weak fields show that the permittivity of $NH_4H_2PO_4$ along the antipolarization directions (a and b axes corresponding to [100] and [010]) and at right-angles to these directions (c axis, corresponding to [001]) obey the Curie−Weiss law above the transition temperature. The polarizabilities of a mechanically free crystal α^t are given by the relationships

$$\alpha_a^t = 0.48 + \frac{1430}{T - (-55)}, \qquad\qquad (VII.94)$$

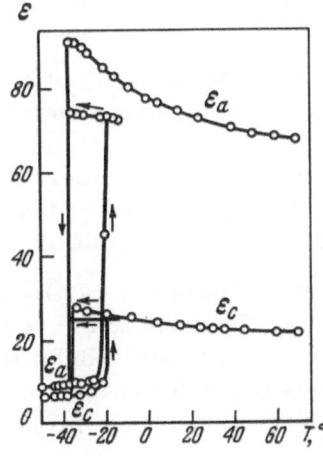

Fig. 201. Temperature dependences of the permittivities ε_a and ε_c of the antiferroelectric $ND_4D_2PO_4$ (along the antipolarization direction) [55].

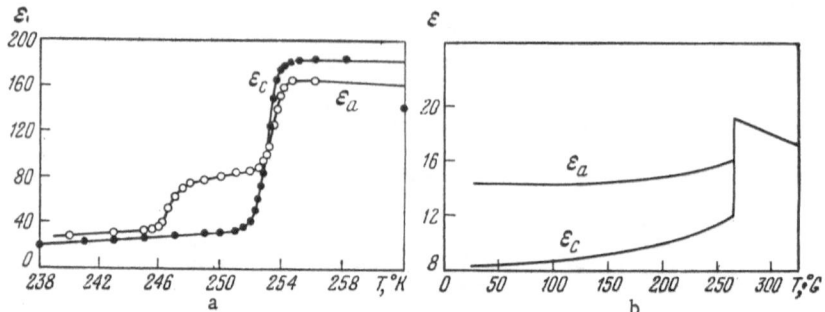

Fig. 202. Temperature dependences of the permittivity of antiferroelectrics $(NH_4)_2H_3IO_6$ (a) [26] and boracite, $Mg_3B_7O_{13}Cl$ (b) [21]. a, b) ε_c along the anti-polarization direction, ε_a at right-angles to this direction.

$$\alpha_c^t = 0.48 + \frac{214}{T-(-17)}, \qquad (VII.95)$$

which indicate that the Curie–Weiss temperatures are negative. Mechanical clamping of these crystals lowers the transition temperature.

The nature of the temperature dependence of the polarization of $NH_4H_2PO_4$-type antiferroelectrics can be deduced from Fig. 201, which shows the temperature dependences of ε_a and ε_c for $ND_4D_2PO_4$. The sudden change in ε at the transition point and the strong thermal hysteresis indicate that $NH_4H_2PO_4$-type antiferroelectrics undergo a phase transition of the first kind. Changes in the structure are so great that these crystals crack when passing through the transition point.

The temperature dependences of the permittivity of antiferroelectrics such as $(NH_4)_2H_3IO_6$ along and at right-angles to the trigonal axis are similar. The fall of ε_a of $(NH_4)_2H_3IO_6$ takes place in two steps, which may indicate the existence of an intermediate phase below the transition point (Fig. 202a). The transition in $Ag_2H_3IO_6$ takes place in a wide temperature range of about 40°C.

There are grounds for assuming that boracite $(Mg_3B_7O_{13}Cl)$, which undergoes a phase transition from the cubic to the ortho-rhombic modification $(3/\overline{4} \rightarrow 2 \cdot m)$ at 265°C, should exhibit ferroelectric or antiferroelectric properties below the phase transition temperature. Fairly careful investigations of the properties

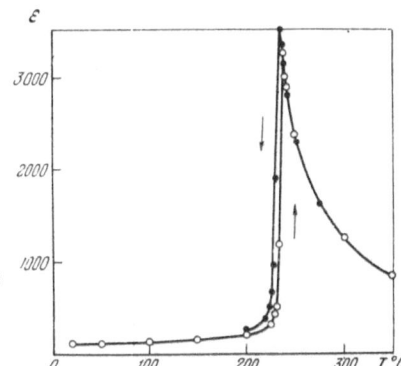

Fig. 203. Temperature dependence of
the permittivity of polycrystalline
PbZrO$_3$ [65].

and structure of this substance, carried out by several workers,
indicate that it is not a ferroelectric but it may be an antiferroelectric. Taking this into account, Fig. 202b shows the temperature dependence of the permittivity of boracite, measured in weak
fields at 500 kc, which can be taken as indicating a phase transition in this crystal.

Among the oxygen-octahedral antiferroelectrics, the compound whose dielectric properties have been studied most thoroughly is PbZrO$_3$. Investigations have shown that the permittivity of
the paraelectric modification of this compound obeys the Curie–
Weiss law. The Curie–Weiss temperature is 190°C, which is lower than the true transition temperature (230°C). In contrast to
other antiferroelectrics, which exhibit ordering of structure elements, the permittivity of PbZrO$_3$ reaches high values (Fig. 203).
The Curie constant of this compound is of the same order of magnitude as that of BaTiO$_3$ (C = 1.59×10^5 deg). The phase transition in PbZrO$_3$ is of the first kind and is characterized by thermal
hysteresis (the temperatures obtained during heating and cooling
differ by 4°).

According to the thermodynamic theory (§5 in Chap. VI), the
application of a constant field lowers the transition temperature of
lead zirconate: a field of 20 kV/cm reduces this temperature by
1.5°. In strong alternating electric fields the polarization of PbZrO$_3$
is described by double hysteresis loops (Fig. 204). The critical
field of a double loop, E_{cr} , is the field in which this antiferroelectric becomes ferroelectric. We note that, in general, double hy-

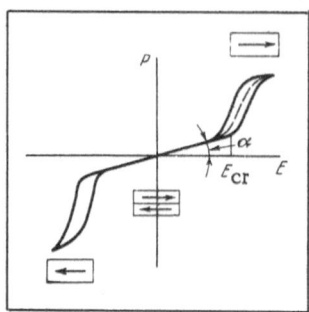

Fig. 204. Double hysteresis loop
of PbZrO₃.

steresis loops are one of the criteria for assigning a substance to
the class of antiferroelectrics. The value of E_{cr} depends on temperature; the closer a crystal is to the transition temperature, the
lower the value of this field. The temperature dependence of E_{cr}
of PbZrO₃ is shown in Fig. 205.

Investigations of the dielectric properties of NaNbO₃ have
shown that only the transition at 360°C is accompanied by a large
change in the permittivity. Transitions at 518 and 640°C are accompanied only by small anomalies of ε. This behavior of NaNbO₃
can be explained by assuming that only the 360°C transition of this
crystal is of ferroelectric nature (a transition between the antiferroelectric and nonpolar phases), whereas the other two transitions are not of ferroelectric nature because NaNbO₃ is not polar
above 360°C.

A characteristic feature of a transition of the first kind, observed at −200°C in NaNbO₃, is that this is a transition between the
ferroelectric state (which exists below −200°C) and the antiferro-

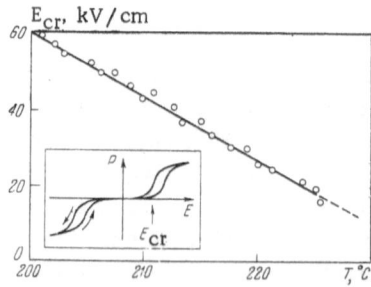

Fig. 205. Temperature dependence of
the critical field E_{cr} of PbZrO₃ [67].
The inset shows the definition of the
critical field.

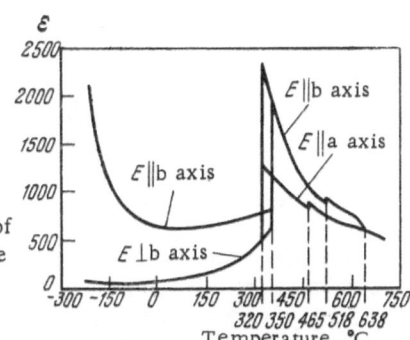

Fig. 206. Temperature dependences of the permittivities of NaNbO$_3$ along the orthorhombic axes [35].

electric modification (which exists between −200°C and +360°C). The spontaneous polarization along the b axis (below −200°C) is about 12×10^{-6} C/cm^2. The spontaneous polarization at right-angles to the b axis is very small, which indicates that domains in this modification cannot be rotated by the field. The temperature dependence of the permittivity of NaNbO$_3$ is shown in Fig. 206.

In Chap. II we have described the structure of NaNbO$_3$ and have mentioned a strong thermal hysteresis of the transition (from −200°C to −10°C). This makes it possible, on the one hand, to use strong fields, perpendicular to the b axis, to observe double hysteresis loops right down to −200°C, and, on the other hand, to produce ordinary hysteresis loops by a single application of a strong polarizing field along the b axis (this produces a transition to the ferroelectric modification at temperatures higher than −200°C).

Information on the nature and special features of the polarization of WO$_3$ is very scarce. All that we know is that this crystal is antiferroelectric in the temperature range 900-740°C. Between 740°C and −50°C this compound is also antiferroelectric and it exhibits a strong electric anisotropy ($\varepsilon_a = 9 \times 10^4$, $\varepsilon_b = 4 \times 10^4$, $\varepsilon_c = 6 \times 10^3$). This result, combined with the x-ray diffraction data, shows that the c axis of this modification of WO$_3$ is the antipolarization axis. The transition at −50°C in WO$_3$ shows a strong thermal hysteresis (−50°C during cooling and −10°C during heating). Below this transition temperature the crystal becomes ferroelectric.

References

1. I. A. Andronova, Izv. Vuzov, Radiofizika, 4:90 (1961).
2. W. F. Brown, Jr., "Dielectrics," Handbuch der Physik (ed. by S. Flügge), Vol. 17, Part 1, Springer Verlag, Berlin (1956), p. 1.
3. A. A. Vorob'ev and E. K. Zavadovskaya, Electric Strength of Solid Dielectrics [in Russian], GTTI, Moscow (1956).
4. V. M. Gurevich and I. S. Zheludev, Izv. Akad. Nauk SSSR, Ser. Fiz., 24:1342 (1960).
5. P. Debye, Polar Molecules, Chemical Catalog Co., New York (1929).
6. I. S. Zheludev, Kristallografiya, 1:105 (1956).
7. I. S. Zheludev, V. V. Gladkii, L. Z. Rusakov, and I. S. Rez, Izv. Akad. Nauk SSSR, Ser. Fiz., 22:1465 (1958).
8. I. S. Zheludev and N. A. Romanyuk, Kristallografiya, 4:710 (1959).
9. I. S. Zheludev and R. Ya. Sit'ko, Kristallografiya, 1:689 (1956).
10. F. Jona and G. Shirane, Ferroelectric Crystals, Pergamon Press, Oxford (1962).
11. W. Känzig, "Ferroelectrics and antiferroelectrics," Solid State Phys., 4:1 (1957).
12. W. J. Merz, Phys. Rev., 111:736 (1958).
13. V. N. Murzin and A. I. Demeshina, Fiz. Tverd. Tela, 6:182 (1964).
14a. V. M. Petrov, Kristallografiya, 6:632 (1961).
14b. V. M. Petrov, Kristallografiya, 7:403 (1962).
15. N. A. Romanyuk and I. S. Zheludev, Kristallografiya, 5:904 (1960).
16. D. G. Sannikov, Zh. Éksp. Teor. Fiz., 41:133 (1961).
17. G. I. Skanavi, Physics of Dielectrics [in Russian], GTTI, Moscow (1949).
18. G. I. Skanavi, Dielectric Polarization and Losses in High-Permittivity Glasses and Ceramics [in Russian], Gosénergoizdat, Moscow (1952).
19. G. I. Skanavi, Physics of Dielectrics in Strong Fields [in Russian], Fizmatgiz, Moscow (1958).
20. A. S. Sonin and V. V. Gladkii, Kristallografiya, 5:145 (1960).
21. A. S. Sonin and I. S. Zheludev, Kristallografiya, 8:57 (1963).
22 B. A. Strukov, V. A. Koptsik, and V. D. Ligasova, Fiz. Tverd. Tela, 4:1334 (1962).
23. H. Frohlich, Theory of Dielectrics, Dielectric Constant and Dielectric Loss, Clarendon Press, Oxford (1958).
24. M. Anliker, H. R. Brugger, and W. Känzig, Helv. Phys. Acta, 27:99 (1954).
25. H. T. Arend and P. Coufová, Czech. J. Phys., 9:751 (1959).
26. P. Bärtschi, Helv. Phys. Acta, 18:267 (1945).
27. T. S. Benedict and J. L. Durand, Phys. Rev., 109:1091 (1958).
28. B. Brežina and A. A. Fotčenkov (Fotchenkov), Czech. J. Phys., B14:21 (1964).
29. B. Brežina and V. Janovec, Czech. J. Phys., B14:44 (1964).
30. G. Busch, Helv. Phys. Acta, 11:269 (1938).
31. A. G. Chynoweth, Phys. Rev., 102:705 (1956).
32. P. Coufová and H. T. Arend, Czech. J. Phys., B10:663 (1960).
33. P. Coufová and H. T. Arend, Czech. J. Phys., B11:416 (1961).
34. P. Coufová and H. T. Arend, Czech. J. Phys., B12:308 (1962).
35. L. E. Cross and B. J. Nicholson, Phil. Mag., 46:453 (1955).
36. M. E. Drougard and R. Landauer, J. Appl. Phys., 30:1663 (1959).

37. V. Dvořák, Czech. J. Phys., 9:710 (1959).
38. E. Fatuzzo, J. Appl. Phys., 32:1571 (1961).
39. E. Fatuzzo and W. J. Merz, Phys. Rev., 116:61 (1959).
40. E. Fatuzzo and W. J. Merz, J. Appl. Phys., 32:1685 (1961).
41. J. Fousek, Czech. J. Phys., 9:172 (1959).
42. J. Fousek and B. Brezina, Czech. J. Phys., B10:511 (1960).
43. P. Glogar and V. Janovec, Czech. J. Phys., B13:261 (1963).
44. J. Hablützel, Helv. Phys. Acta, 12:489 (1939).
45. G. G. Harman, Phys. Rev., 111:27 (1958).
46. S. Hoshino, T. Mitsui, F. Jona, and R. Pepinsky, Phys. Rev., 107:1255 (1957).
47. J. K. Hulm, Phys. Rev., 92:504 (1953).
48. V. Janovec, Czech. J. Phys., 8:3 (1958).
49. V. Janovec, Czech. J. Phys., 9:468 (1959).
50. V. Janovec, B. Brezina, and A. Glanc, Czech. J. Phys., B10:690 (1960).
51. W. Känzig, Phys. Rev., 98:549 (1955).
52. C. Kittel, Phys. Rev., 83:458 (1951).
53. R. Landauer, D. R. Young, and M. E. Drougard, J. Appl. Phys., 27:752 (1956).
54. E. A. Little, Massachusetts Inst. Technol., Report No. 87 (1954).
55. W. P. Mason and B. T. Matthias, Phys. Rev., 88:477 (1952).
56. B. T. Matthias, Science, 113:591 (1951).
57. W. J. Merz, Phys. Rev., 76:1221 (1949).
58. W. J. Merz, Phys. Rev., 95:690 (1954).
59. W. J. Merz, J. Appl. Phys., 27:938 (1956).
60. W. J. Merz, "Ferroelectricity," Progress in Dielectrics, 4:101 (1962).
61a. R. C. Miller and A. Savage, Phys. Rev. Letters, 2:294 (1959).
61b. R. C. Miller and A. Savage, Phys. Rev., 112:755 (1958).
61c. R. C. Miller and A. Savage, Phys. Rev., 115:1176 (1959).
61d. R. C. Miller and A. Savage, J. Appl. Phys., 31:662 (1960).
62. T. Mitsui and J. Furuichi, Phys. Rev., 90:193 (1953).
63. L. Pauling, Proc. Roc. Soc. (London), A114:181 (1927).
64. C. F. Pulvari and W. Kuebler, J. Appl. Phys., 29:1315, 1742 (1958).
65. S. Roberts, J. Am. Ceram. Soc., 33:63 (1950).
66. S. Sawada, S. Nomura, and Y. Asao, J. Phys. Soc Japan, 16:2207 (1961).
67. E. Sawaguchi and K. Kittaka, J. Phys. Soc. Japan, 7:336 (1952).
68. C. C. Stephenson and H. E. Adams, J. Am. Chem.Soc., 66:1409 (1944).
69. C. C. Stephenson and A. C. Zettlemoyer, J. Am. Chem. Soc., 66:1405 (1944).
70. E. C. Subbarao, M. C. McQuarrie, and W. R. Buessem, J. Appl. Phys., 28:1194 (1957).
71a. B. Szigeti, Trans. Faraday Soc., 45:155 (1949).
71b. B. Szigeti, Proc. Roy. Soc. (London), A204:51 (1950).
72. J. R. Tessman, A. H. Kahn, and W. Shockley, Phys. Rev., 92:890 (1953).
73. S. Triebwasser, IBM J. Res. Developm, 2:212 (1958).
74. H. H. Wieder, J. Appl. Phys., 26:1479 (1955).
75. H. H. Wieder, Phys. Rev., 110:29 (1958).

Chapter VIII

Electrical Conduction and Dielectric Losses

Introduction

Electrical conduction in dielectrics is the ordered motion of weakly bound charged particles under the influence of an electric field. The available information on electrical conduction in solid dielectrics refers mainly to polycrystalline and amorphous materials; electrical conduction in single-crystal dielectrics has not yet been investigated sufficiently thoroughly.

The conduction process is measured in terms of an important parameter called the electrical conductivity. Studies of the electrical conductivity make it possible to determine the structure of a crystal and to obtain information on the lattice energy, as well as on defects and impurities, etc.

Crystals of linear dielectrics and ferroelectrics exhibit mainly ionic conduction. Pronounced electronic conduction is observed only in crystals of oxygen-octahedral ferroelectrics ($BaTiO_3$, etc.), which is due to the presence of oxygen defects in the structure of these crystals. Consequently, we shall concentrate in this chapter on the ionic conduction both of the intrinsic and extrinsic (impurity) type. We shall consider in detail the mechanisms of interstitial conduction (in the presence of Frenkel defects) and vacancy conduction (in the presence of Schottky defects). We shall also discuss in detail the processes of the formation of space charge, the mechanism of electrical "forming," the concepts of "true" and residual resistance, the nature of high-voltage polarization, etc.

The special nature of conduction in ferroelectrics is associated with the structure and phase transitions in these crystals.
These concepts will be stressed in the description of ferroelectrics.

Dielectric losses are the dissipation of the electric field energy in dielectrics. They are closely related to the processes of
electric polarization and conduction in dielectrics. Unfortunately,
like electric conduction, dielectric losses in singly-crystal materials have not been studied in sufficient detail. This is particularly true of ferroelectric crystals.

In the last two sections of this chapter, we shall describe the
physics of dielectric losses and give the rudiments of the theory of
losses in solid dielectrics. Attention will be concentrated on losses
associated with the establishment of the thermal types of polarization: thermal orientational and thermal ionic polarization.

The special nature of dielectric losses in ferroelectrics is
due to the presence of domains. In weak fields, the contribution of
domain processes to the polarization is frequently small and therefore the losses in such crystals are fairly small. In strong fields,
the polarization reversal (switching) processes in ferroelectrics
are accompanied by very large losses due to hysteresis. It is also worth mentioning here the dielectric losses in ferroelectrics in
the millimeter range of wavelengths, where dielectric relaxation
associated with lattice vibrations is observed. Unfortunately, little
has been published on dielectric losses in ferroelectric crystals.

Major contributions to our knowledge of electric conduction
and dielectric losses in general and in solid dielectrics in particular have been made by Soviet scientists, including A. I. Ioffe,
Ya. I. Frenkel' (J. Frenkel), G. I. Skanavi, and others. A detailed
exposition of the knowledge available up to 1952 can be found in
Skanavi's books [16, 17]. The treatment of the topics in §§1, 2,
and 4 is based, to a large extent, on Skanavi's books.

§1. Ionic Conduction in Crystals

Ionic conduction differs from the electronic process by the
transport of matter which is known as electrolysis. Dielectrics
exhibiting ionic conduction should satisfy Faraday's law

$$Q = kIt,$$

(VIII.1)

where Q is the amount of electric charge transferred; I is the current; k is the electrochemical equivalent; t is time.

In dielectrics in which conduction is solely due to ions (of one or both signs) the nature of conduction (i.e., whether they satisfy Faraday's law) can be established by weighing three cylinders before and after the application of a field (after electrolysis). If the conduction is due to ions of one sign the application of a field reduces the weight of the first of the three cylinders placed in series, increases the weight of the last one, but does not affect the weight of the middle one. Knowing the direction of the field and the amount of charge that flows during the experiment, we can easily determine the sign of the ions carrying the charge as well as the value of the electrochemical equivalent k. An exact knowledge of the change in weight, the value of k, and the sign of charge carriers can sometimes be used to determine even the actual ion which carries the current in a given crystal.

Investigations of the nature of conduction by the method of three cylinders have been carried out on a range of crystals. They show that the principal charge carriers are those ions which have the smallest size for a given charge or those ions which have a smaller charge for a similar size. For example, the main charge carriers in NaCl are the Na^+ ions; in $PbCl_2$ they are the Cl^- ions, etc. (Table 42). The presence of small monovalent ions always increases the electrical conductivity of a crystal.

In some crystals the process of conduction is due to ions of two signs (for example, in PbI_2). Moreover, when the temperature is varied we find that, in some crystals, ionic charge carriers of one sign are joined by carriers of opposite sign. Thus, for example, above 600°C chlorine ions join (although to a small degree) sodium ions in charge transport. The nature of conduc-

TABLE 42. Charge Carriers in Various Crystals [16]

Crystal	Temperature interval, °C	Charge carrier	Crystal	Temperature interval, °C	Charge carrier
AgCl	20—350	Ag^+	BaF_2	to 500	F^-
AgBr	20—300	Ag^+	$BaBr_2$	350—450	Br^-
AgI	20—400	Ag^+	$PbCl_2$	200—450	Cl^-
NaF	below 500	Na^+	$TlBr_2$	200—300	Br^-
NaCl	below 400	Na^+			

tion and the type of ion in the case of charge transport by ions of two signs can also be established, in principle, by the method of weighing the cylinders but, in this case, the problem is more complicated and it is necessary to use additional cylinders in which the charge carrier is known, etc.

Some crystals exhibit a mixed electronic–ionic type of conduction. Difficulties are encountered in the studies of such crystals since the results may apparently indicate that Faraday's law is satisfied. This has been found for crystals of α-Ag_2S, which have predominantly electronic conduction. Examples of crystals exhibiting mixed conduction are also Ag_2Se and Ag_2Tl crystals. The predominantly electronic nature of the conduction in Ag_2S, Ag_2Se, and Ag_2Tl can be established using the Hall effect. These crystals are basically semiconductors. We must point out, however, that not all crystals exhibiting purely ionic conduction are good dielectrics (insulators).

In strong fields the electronic conduction is frequently superimposed on the ionic process observed in many crystalline dielectrics. This effect (deviation of the conduction process from Faraday's law) has been found in quartz crystals, in rocksalt, and in other materials. Thus in strong fields and at low temperatures conduction in mica is mainly electronic but in weak fields and at high temperatures the dominant process is ionic.

A. Ionic Conduction Mechanism. Irrespective of the actual mechanism of conduction, we can establish general relationships between the current (which appears when charged particles move under the action of an electric field) and the concentration of charged particles. Let us assume that the number of such particles per 1 cm^3, capable of motion under the action of a field, is equal to n, and the charge of each particle is e. The force exerted on each particle is eE, where E is the electric field intensity. The average velocity of the particle v is proportional to the field:

$$v = \varkappa E, \qquad\qquad\qquad (VIII.2)$$

where $\varkappa$ is the mobility of charged particles. The application of a field to a dielectric thus produces a current of density j (which represents the amount of charge crossing 1 cm^2 per unit time):

$$j = nev = ne\varkappa E. \qquad \text{(VIII.3)}$$

Using the relationship between the current density, field intensity, and electrical conductivity σ,* and assuming that Ohm's law applies, we have

$$\sigma = \frac{j}{E}, \qquad \text{(VIII.4)}$$

and it follows from (VIII.3) that

$$\sigma = ne\varkappa. \qquad \text{(VIII.5)}$$

The density of a current in a dielectric can be expressed also in terms of the mean free path of a particle δ. In this approach it is assumed that the current in a dielectric is due to the motion of various particles which are displaced by a distance δ; when one particle stops, it is replaced by another, etc. If the number of particles activated per 1 cm^3 of a dielectric in 1 sec is N, we find that a given area is crossed in 1 sec only by those charged particles which are liberated at points separated from a given area by distances not greater than δ (it is assumed that the mean free path δ is much less than the path which can be traversed by a charge in one second). Thus,

$$j = Ne\delta. \qquad \text{(VIII.6)}$$

It must be stressed that the current density in a dielectric is governed by the average rate of displacement of charged particles along the direction of the field and not by the instantaneous velocities of individual particles; the mobility of a particle is determined using its mean free path. Thus, these two methods of describing the electrical conductivity (in terms of the mobility $\varkappa$ or mean free path δ) are equivalent.

In ionic crystals the process of electrical conduction may be due to the motion of ions of the host crystal lattice. Such conduction (and the corresponding conductivity) is known as intrinsic

---

*The electrical conductivity σ (like the resistivity ρ) is not a scalar in an anisotropic medium but a polar tensor of the second rank. Thus, "geometrical" features of the electrical conductivity are similar to those observed in polarization processes (§1, Chap. VII).

and it is particularly important at high temperatures. On the other hand, electrical conduction in ionic crystals may be due to the motion of relatively weakly bound ions. They include impurity ions as well as ions located at defects in the crystal lattice. These ions give rise to extrinsic or impurity conduction (and extrinsic or impurity conductivity) of a crystal. Conduction associated with the motion of weakly bound ions may appear at relatively low temperatures. In many cases, the same crystal exhibits electrical conduction due to host ions as well as weakly bound ions. Non-ionic crystals (for example, molecular crystals) usually exhibit extrinsic conduction.

It must be stressed that ionic conduction is closely associated with lattice defects and not all the ions present in a crystal take part in the conduction process.

Major contributions to our knowledge of electrical conduction in ionic crystals have been made by Ioffe, Frenkel, and Schottky.

Let us consider a crystal AB, for example, a halide of a monovalent metal or a sulfide (oxide) of an alkaline-earth metal. If a crystal is perfect in its equilibrium state the volume (bulk) ionic conduction is observed only when positive or negative ions leave their normal positions and can move through the lattice under the action of a field. In this case, defects are generated by the field. However, in order to detach an ion from its normal position the field must be very high. In fact, the potential energy of an ion varies by a value of the order of 1 eV over an interatomic distance and this means that a current in a dielectric which is solely due to the application of a field would flow only in fields of the order of 10^6 V/cm. Hence, we must assume that charge-carrying ions are displaced before the application of a field and that the field simply orders their random motion.

The motion of ions through a crystal can take place in two ways: a) ions can move between lattice sites (in a perfect crystal the interstitial positions are free and they are occupied only in crystals containing interstitial atoms); b) ions can move by jumping to unoccupied sites which are known as vacancies or holes. Such motion of ions is sometimes regarded as the motion of vacancies.

In order to obtain the equation describing the process of electrical conduction due to interstitial ions (case a) or vacancies (case b), we shall first consider the probability w that an ion (or vacancy) overcomes (by thermal motion in the absence of an electric field) a barrier of height $\mathscr{E}_s^-$ along a direction α:

$$w = \frac{1}{\alpha} \frac{(kT)^3}{h^3 v_s^2} e^{-\frac{\mathscr{E}_s}{kT}}. \qquad (VIII.7)$$

We shall next consider the total probability $\tilde{w}_p$ that an ion overcomes a potential barrier under the action of an electric field E, which deforms the potential barrier. In weak fields (Eex $\ll$ kT, where x is the distance between two equilibrium positions) the total probability of an ion jump along the direction of the field is

$$\tilde{w}_p = w \frac{Eex}{kT}. \qquad (VIII.8)$$

Since the electric polarization associated with each successful jump is ex, we can use Eq. (VIII.8) to find the expression for the density of the current

$$j = nw \frac{Ee^2 x^2}{kT}, \qquad (VIII.9)$$

where n is the number of interstitial ions or vacancies per unit volume. This makes it possible to find the conductivity σ due to ions or vacancies:

$$\sigma = nw \frac{e^2 x^2}{kT}. \qquad (VIII.10)$$

Using the relationship between the conductivity σ and the mobility $\varkappa$ of Eq. (VIII.5) we can calculate the mobility of ions or vacancies:

$$\varkappa = w \frac{ex^2}{kT}. \qquad (VIII.11)$$

In order to estimate the values of σ and $\varkappa$ in specific cases, we must calculate the value of w. In the case of interstitial conduction this quantity is the probability of a thermal jump of an ion from one interstice to another, whereas in the case of vacancy conduction, it is the probability of a jump of an ion from one va-

cancy to another. In order to obtain a qualitative idea of which conduction mechanism takes place in a particular case, it is necessary to take into account the relationship between the energy of an
ion in its bound state (at a lattice site) U_1, its energy at an interstice (metastable state) U_2, and the height of the potential barrier
separating two metastable states, U_0.

If the energy of formation of a vacancy in a crystal, E_2, is
higher than the energy necessary for the transfer of an ion to a
metastable state, $E_1 = U_2 - U_1$, in a loosely packed crystal lattice,
the probability of formation of an interstitial ion will be greater
and the current in a crystal will be due to the motion of ions from
one metastable position to another, overcoming barriers of height
U_0 on the way. Such interstitial conduction may be called the excess conduction (in contrast to the vacancy or hole conduction).
The value and temperature dependence of the interstitial conductivity depends on the relationship between the total number of ions
and the number of sites at which ions may be bound. This relationship can be considered in two limiting cases: 1) the concentrations of free interstices and vacancies in a lattice are high compared with the total number of ions; 2) the total number of ions is
equal to the number of sites, i.e., a site becomes free (a vacancy
is formed) only when an ion is transferred to an interstice (it is
still assumed that the number of possible metastable states of ions
at interstices is high). The second case corresponds to a crystal
with Frenkel defects (Fig. 207a).

In both limiting cases the electrical conductivity can be
written in the form

$$\sigma = A e^{-\frac{B}{T}}, \qquad\qquad (VIII.12)$$

on condition that in the first case $B = E_0/k = (E_1 + U_0)/k$, where E_0
is the total energy required for the activation of an ion in its motion in the interstitial space, and that in the second case we have
$B = (E_0 + U_0)/2k$.

In the presence of Schottky defects (Fig. 207b), when the energy of formation of a vacancy E_2 is less than the energy of formation of an interstitial ion E_1, we should observe vacancy conduction.
An analysis shows that once again the electrical conductivity is described by Eq. (VIII.12), provided that $B = (E_0 + 0.5\,E_2)/k$.

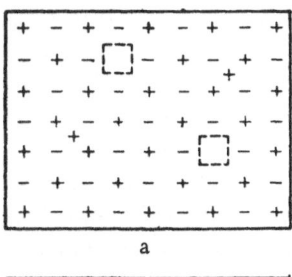

a

Fig. 207. Frenkel (a) and
Schottky (b) defects in ionic
crystals.

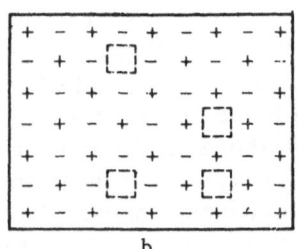

b

The same temperature dependence of the electrical conductivity, Eq. (VIII.12), applies to the interstitial and vacancy conduction mechanisms; it is assumed that B in that equation is completely independent of temperature and that A varies weakly with
temperature.

If the conduction process is due to ions of different types,
Eq. (VIII.12) transforms to

$$\sigma = \sum A_i e^{-\frac{B_i}{T}}, \qquad \text{(VIII.13)}$$

where A_i and B_i are characteristics of each of the mobile ions
(vacancies).

We shall now make some comments on the intrinsic conductivity of crystals. Theoretical considerations cannot be used to estimate quantitatively (with sufficient accuracy) the relationships between the energies E_1 and E_2 and thus establish the conduction mechanism in any specific case. However, calculations can be used
to obtain some results. In particular, the energy of an ion at an
interstice in an alkali halide crystal is close to zero. This indicates that the probability of finding an ion at an interstice in an alkali halide crystal is low. Moreover, the difference between the

energy of a completely filled lattice of an alkali halide crystal and a lattice with vacancies is considerably smaller than the difference between the energy of a completely filled lattice and a lattice with ions at interstices. This again indicates that the presence of vacancies in alkali halide crystals is much more likely than the presence of ions at interstitial positions. Thus, the motion of ions in alkali halides should consist mainly of jumps of ions from one site to any neighboring vacant site (vacancy conduction). On the other hand, the motion of ions along interstices should be expected in crystals with loosely packed lattices consisting of ions differing greatly in size. This applies, for example, to AgBr crystals in which the ionic radius of silver is 1.13 Å and that of bromine is 1.95 Å. In these crystals a silver ion can easily travel from a tetrahedral cavity (which is its main position in the lattice) to an octahedral cavity and, moving from one octahedral cavity to another under the action of a field, it can give rise to electrical conduction.

General considerations show clearly that ionic conduction in crystals is mainly due to the motion of smaller ions, i.e., usually positive ions (cations).

The absolute values of the energies required for the transfer of ions to various positions can be estimated from the expression $(0.1-0.5) \, e^2/r_0$, where r_0 is the ionic radius; this expression gives energies of the order of $(0.8-4) \times 10^{-12}$ erg or 0.5-2.5 eV. The absolute value of the coefficient A in Eq. (VIII.12) is of the order of $10^6 \, \Omega^{-1} \cdot cm^{-1}$ for alkali halide crystals.

If the electrical conduction in a crystal is due to impurities (this is frequently encountered in covalent crystals), these impurity ions are usually attached to lattice defects provided, of course, that the number of such defects exceeds the number of impurity ions (this corresponds to the first case of the interstitial conduction described above.

We have already mentioned that in a real crystal conduction at high temperatures is mainly intrinsic but at low temperatures it is primarily extrinsic (associated with impurities or defects). This makes it necessary to describe the temperature dependence of the electrical conductivity by an expression of the type

$$\sigma = A_1 e^{-\frac{B_1}{T}} + A_2 e^{-\frac{B_2}{T}}, \qquad (VIII.14)$$

where the subscripts "1" of A and B refer to the host ions of the
lattice and the subscripts "2" refer to impurity ions (Fig. 208).
We must bear in mind that the number of foreign impurity ions in
high-quality crystals is very small compared with the number of
host lattice ions.

The presence of impurities affects in different ways the low-
temperature interstitial and vacancy conduction mechanisms. Thus,
if conduction in a crystal is due to the motion of ions along interstices, the introduction of impurity ions should increase the electrical conductivity very strongly since, in this case, the number
of vacant sites in the lattice is small and impurity ions enter the
interstices. This increases the number of charge carriers and
the distortion of the lattice so that the activation energies of impurities become small. This corresponds to the second case of
the interstitial conduction mechanism. Conversely, in the vacancy conduction case, the introduction of a small number of impurities should not increase the conductivity very greatly. In this case,
impurities can occupy existing vacancies; this produces practically no distortion of the lattice and the reduction in the number of
vacancies may even reduce the conductivity. Finally, impurities
can also move in defective regions of a crystal, corresponding to
the first case of the interstitial conduction mechanism. In general, the role of impurities in electrical conduction is not simple
and all that we can say is that the electrical conductivity of a crystal is governed, in the final analysis, by the concentration of defects, the mobility of vacancies, the concentration of ions at interstices, and the concentration of impurity ions.

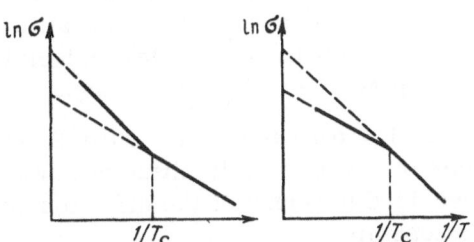

Fig. 208. Schematic representations of the dependence $\ln \sigma = f(1/T)$. T_c is the temperature
corresponding to a change in the nature of conduction.

A change in the slope of the dependence of ln σ on $1/T$, observed for some crystals (Fig. 208), may be associated not only with the intrinsic and interstitial conduction mechanisms but also with other factors. In particular, this change in the slope may be explained by assuming that the coefficient B in Eq. (VIII.12) decreases with increasing temperature. The rate at which an equilibrium number of charge carriers is reached at a given temperature T' may become so slow that the same equilibrium state is retained also at lower temperatures. The temperature T' then corresponds to the temperature of a kink in the straight lines representing the dependence ln $\sigma = f(1/T)$. We note that in the case of alkali halide crystals the slope of the straight line in the low-temperature part is approximately one-third of the slope in the high-temperature part.

In the introduction to this section we have mentioned that when the temperature is varied the nature of the intrinsic conduction of a crystal can change as well. The possibility of the activation of ions at high temperatures (cations are activated at lower temperatures) was first pointed out by A. F. Ioffe. There is a point of view according to which the kink in the dependence ln $\sigma = f(1/T)$, of the type shown in Fig. 208 and observed in alkali halide crystals, is due to the fact that at low temperatures the conduction is due to alkali metal ions but at high temperatures halogen ions take part in charge transport. Other investigators are of the opinion that in purely ionic crystals (particularly in NaCl) the high-temperature electrical conductivity is governed by the number of interstitial ions (Frenkel defects) and the low-temperature conductivity is governed by the number of vacancies (Schottky defects). We must also mention that a change in the slope of the straight lines ln $\sigma = f(1/T)$ may be due to a phase transition in a crystal. This is exhibited particularly clearly by ferroelectric crystals (§3 in the present chapter).

B. Electrical Conductivity and Self-Diffusion. The motion of atoms in crystals (self-diffusion, heterodiffusion, etc.) is usually described in terms of the diffusion coefficient D, defined by the relationship

$$D = D_0 e^{-\frac{E}{kT}}. \qquad \text{(VIII.15)}$$

The diffusion coefficient D has the dimensions of cm^2/sec and is equal to the mass of matter (measured in gram-molecules) which diffuses in 1 sec through an area of 1 cm^2 when the concentration gradient is 1 mole/cm; E is the activation energy of diffusion, equal to the energy necessary for the transfer of an atom from one position to another; D_0 is the pre-exponential factor.

The diffusion coefficient D varies with temperature in accordance with the same exponential law as the ionic conductivity, and the constants in the two relations are the same [14]. The similarity of the diffusion and ionic conduction phenomena makes it possible to take into account the displacements of ions which cannot be included in the determination of the transport (transference) numbers in measurements of the ionic electrical conductivity.

The diffusion coefficient D and the ionic mobility $\varkappa$ in an electric field are related by the Einstein equation:

$$\frac{\varkappa}{D} = \frac{e}{kT} ; \qquad\qquad\qquad (VIII.16)$$

using Eq. (VIII.15), we can find the relationship between the electrical conductivity σ and the diffusion coefficient D

$$\frac{\sigma}{D} = \frac{ne^2}{kT} , \qquad\qquad\qquad (VIII.17)$$

where n is the number of pairs per unit volume. The mobilities of ions, determined from the Einstein equation and from measurements of the electrical conductivity, do not always agree. This is because not all the diffusion processes give rise to an electrical current. In particular, two neighboring vacancies moving due to a concentration gradient (diffusion) do not affect the electrical conductivity. Moreover, in some crystals neighboring ions or atoms can exchange places. Again such diffusion is not accompanied by an electric current.

The agreement between the values of the activation energy, found by the diffusion and electrical conductivity methods, suggests the same mechanism for the motion of ions in diffusion and in the flow of an electric current in an ionic crystal. As a rule, the values of the mobility (or diffusion coefficient) calculated from

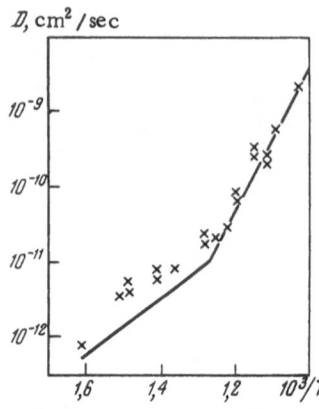

Fig. 209. Temperature dependence of the self-diffusion coefficient D of sodium ions in NaCl.

the Einstein equation are in agreement with the experimental values. Figure 209 gives, by way of example, the temperature dependence of the self-diffusion coefficient of radioactive sodium ions in NaCl crystals. The continuous line is drawn using the values of the self-diffusion coefficient calculated from the Einstein equation. The crosses represent the experimental values of the self-diffusion coefficient. We can see from this figure that, below 500°C, the measured values of the self-diffusion coefficient are more than twice as large as the calculated values. This is because, at low temperatures, the electrical conductivity is due to structure defects. At high temperatures, in the intrinsic conduction region, the measured and calculated values of the self-diffusion coefficient are in satisfactory agreement.

The difference between the calculated and measured values of the diffusion coefficients may sometimes be due to the difference between the permittivity of the medium and the permittivity of the charge carriers. This makes it possible to use the differences between the calculated and experimental values of D to draw some conclusions about the nature of charge carriers. Summarizing subsections A and B of this section, and before going over to discussion of the experimental investigations of the electrical conductivity, we must stress that, in spite of the fact that the electrical conductivity cannot be calculated directly by theoretical methods (because of difficulties in the calculation of the activation energy), the modern theory makes it possible to describe correctly and to understand the basic features of the mechanism of electrical conduction in crystals.

C. Some Results of Experimental Investigations of Ionic Electrical Conductivity. Experiments show that the high-temperature activation energy decreases with increasing anion radius (when the cation is the same), as shown in Fig. 210. This is because the polarizability, which increases with increasing anion size, tends to increase the cation mobility. The polarizability of cations increases along the sequence Pb, K, Na, and Li. When the cation radius increases (for the same anion), the activation energy increases.

The activation energy at low and high temperatures depends on the lattice energy of a crystal: for the same cation in a series of alkali halide compounds the activation energy increases with increasing lattice energy (from I to Cl, and from Cl to F). Replacement of a cation is accompanied by a transition to a different dependence of the activation energy on the lattice energy. The activation energy of electrical conduction in alkali halides thus shows a periodic variation (Fig. 211). Other properties of crystals can also be related to their electrical conductivity. In particular the electrical conductivity of crystals of one series, for example chlorides, increases with increasing lattice constant.

These relationships governing the activation energy influence the electrical conductivity. The high-temperature intrinsic electrical conductivity of alkali halides thus varies periodically with the composition. An increase of the lattice energy of a series of compounds with the same cation is accompanied by a decrease of the electrical conductivity; when a cation is changed, we have a

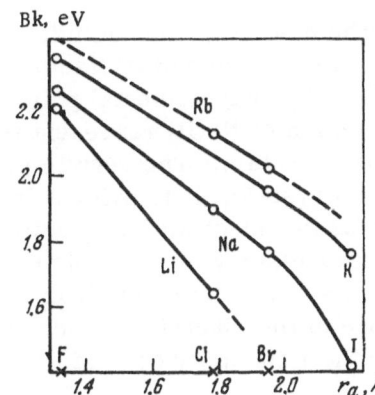

Fig. 210. Dependence of the activation energy on the radii of anions r_a in alkali halides.

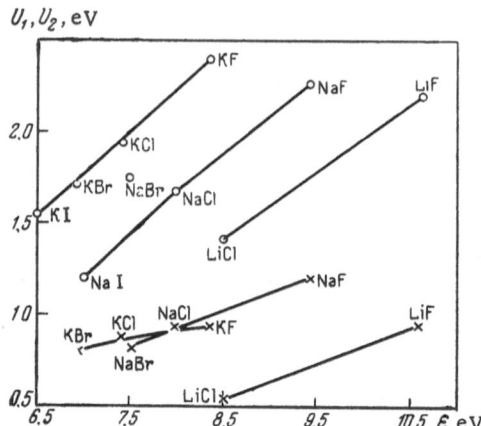

Fig. 211. Dependences of the low-temperature U_2 and high-temperature U_1 activation energies (denoted by crosses and circles, respectively) on the lattice energy of alkali halides.

different monotonic dependence of the electrical conductivity on the lattice energy. Replacement of the lithium cation with sodium, potassium, or rubidium cations, i.e., an increase of the cation radius, reduces the electrical conductivity at a given temperature. This shows that cations play an active role in the transport of charge in alkali halides. The decrease in the electrical conductivity as the cation radius gets larger is probably because the larger ions move less easily through the lattice. The electrical conductivity decreases with increasing number of ions per 1 cm^3 of the lattice.

The electrical conductivity of alkali halide crystals (Fig. 212) can be described satisfactorily by the two-term formula (VIII.14). The experimental points for each of the halides (with the exception of NaCl) represent the results obtained using two synthetic crystals. The results given for NaCl refer to natural rocksalt crystals. It is evident from Fig. 212 that at low temperatures the electrical conductivity of alkali halides depends on the earlier history of a sample and on the impurity concentration. This is particularly clear in Fig. 213, which shows the dependence of the electrical conductivity of potassium chloride crystals on their impurity content.

Most of the impurities present in a crystal are captured during growth. In some cases, the uniformity of a crystal and its impurity distribution depend on the method of growth (for example, they may be different for crystals grown from solution and from the melt). Repeated recrystallization reduces the concentration of impurities and, in some cases (as demonstrated by Ioffe in the case of potash alum), it may reduce the electrical conductivity. The electrical conductivity of recrystallized samples has been found by Ioffe to be tens or even hundreds of times lower than the conductivity of freshly grown crystals. He has found that after

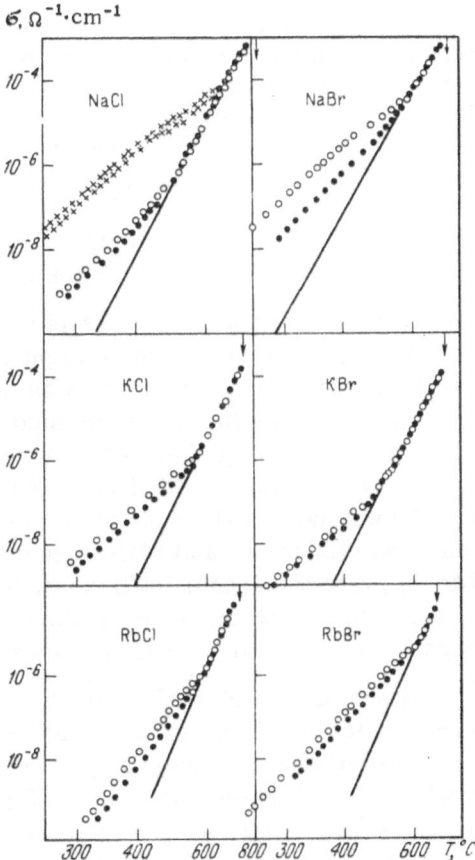

Fig. 212. Temperature dependences of the electrical conductivity of some alkali halides.

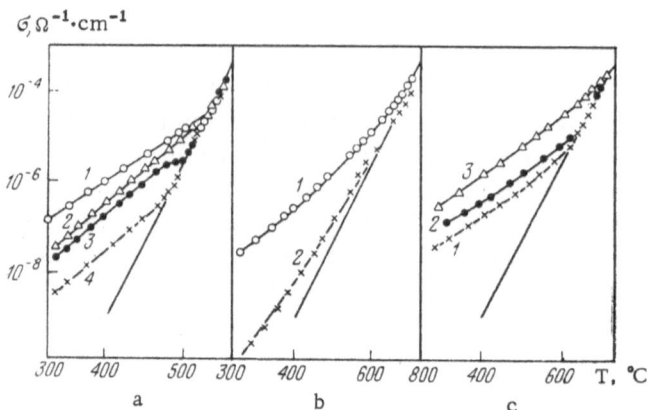

Fig. 213. Influence of impurities on the electrical conductivity of KCl crystals. a: 1) 0.07 mol. % Pb; 2) 0.4 mol. % Pb; 3) natural sylvite; 4) 0.2 mol. % Cu and 0.2 mol. % Ag. b: 1) Before electrical refining; 2) after refining. c: 1) initial material; 2) 0.3 mol. % Cu; 3) 0.3 mol. % Cu and 0.3 mol. % Pb.

triple recrystallization the value of the electrical conductivity is no longer affected by additional recrystallization treatments.

The experimentally determined high-temperature electrical conductivities of alkali halides fit well the straight lines $\ln \sigma = f(1/T)$, irrespective of the earlier history of a sample, which shows that at these temperatures the intrinsic conduction mechanism is dominant. The temperature dependences of the electrical conductivity of silver and thallium halides are similar to those shown in Fig. 212. A summary of the data on the electrical conductivity of some ionic crystals at low and high temperatures is given in Table 43. Table 44 gives the transport (transference) numbers n_+ and n_-, which represent the components of the current carried by positive and negative ions, respectively, in some ionic crystals.

At low temperatures the electrical conductivity of alkali halides is due to impurities or is due to host ions associated with defective regions in the lattice; at these temperatures the value of A_1 is considerably higher than the value of A_2 [Eq. (VIII.14)] and the ratio A_1/A_2 reaches 10^6. As already mentioned, considerable difficulties are encountered in calculations of the activation energy of charge carriers in crystals exhibiting ionic conduction. We must mention here that there is as yet no rigorous theoretical

TABLE 43. Electrical Conduction and Other Parameters of Some Ionic Crystals [4, 16]

Crystal	Melting point, °C	Lattice energy per ion pair, eV	Ionic radius, Å		Activation energy Bk, eV	
			cations	anions	low-temperature B_2k	high-temperature B_1k
LiF	842	10.56	0.78	1.33	0.91	2.20
NaF	992	9.46	0.98	1.33	1.20	2.25
KF	846	8.37	1.33	1.33	0.91	2.35
RbF	—	8.06	1.49	1.33	—	—
LiCl	606	8.50	0.78	1.81	0.58	1.41—1.65
NaCl	800	7.93	0.98	1.81	0.89	1.72—2.60
KCl	768	7.23	1.33	1.81	0.88	1.90—2.06
RbCl	712	6.95	1.49	1.81	1.35	2.12
LiBr	—	8.05	0.78	1.96	—	1.78—1.90
NaBr	735	7.55	0.98	1.96	0.86	1.83—1.97
KBr	723	6.94	1.33	1.96	0.74	2.03
RbBr	680	6.71	1.49	1.96	0.97	—
LiI	—	7.51	0.78	2.20	—	1.22—1.23
NaI	661	7.07	0.98	2.20	—	1.61—1.77
KI	680	6.55	1.33	2.20	1.00	—
RbI	—	6.34	1.49	2.20	—	—

proof that the activation energy of host ions in ionic crystals is indeed close to 2-2.5 eV.

We shall now make some comments on the electrical conductivity of some nonionic crystals. The electrical conductivity of such crystals is usually described satisfactorily by a formula

TABLE 44. Transport Numbers of Ionic Crystals at Various Temperatures [11]

Compound	Material, °C	Transport numbers		Compound	Material, °C	Transport numbers	
		n_+	n_-			n_+	n_-
NaF	500	1.000	0.000	KCl	600	0.884	0.116
	550	0.966	0.004	AgCl	20—350	1.00	0.00
	600	0.916	0.084	AgBr	20—300	1.00	0.00
	625	0.861	0.139	BaF_2	500	0.00	1.00
NaCl	400	1.000	0.000	$BaCl_2$	400—700	0.00	1.00
	510	0.981	0.019	PbF_2	200	0.00	1.00
	600	0.946	0.054	$PbCl_2$	200—450	0.00	1.00
	625	0.929	0.071	$PbBr_2$	250—355	0.00	1.00
KCl	435	0.956	0.044	PbI_2	255	0.39	0.61
	500	0.941	0.059		290	0.67	0.33
	550	0.917	0.083				

of the type

$$\ln \sigma = A - \frac{B}{T} \qquad\qquad\qquad \text{(VIII.18)}$$

and it is usually due to impurity ions. In anisotropic crystals of quartz the electrical conductivity along the c axis (the optic axis) is higher than at right-angles to this direction (the activation energies are 0.88 and 1.32 eV, respectively). The room-temperature resistivity of quartz along the optic axis is about 10^{14} $\Omega \cdot$cm and at right-angles to this axis it is about 10^{16} $\Omega \cdot$cm. At 500°C the resistivity of quartz falls by five orders of magnitude, compared with the room-temperature value, and becomes about 10^9 $\Omega \cdot$cm along the c axis and 10^{11} $\Omega \cdot$cm at right-angles to this axis. It is very likely that the charge carriers in quartz are Na^+, K^+, and Li^+ impurity ions.

The electrical conductivity of quartz may be altered by electrolytic refining (this is done by the application of a constant electric field during heating), and it depends on the previous history of a sample, the electrode material, and other factors. The dependence $\ln \sigma = f(1/T)$ has a kink at the $\beta - \alpha$ transition temperature (573°C).

The resistivity of synthetic quartz is usually two or three orders of magnitude higher than that of natural quartz. Moreover, since synthetic quartz frequently contains inclusions of the mother liquor captured during growth, the dependence $\ln \sigma = f(1/T)$ has a kink in the temperature range 250–450°C. The activation energy of carriers in this temperature range is lower than at other temperatures. This special characteristic of the conductivity of synthetic quartz disappears after electrolytic refining. The conductivity of synthetic quartz is also likely to be due to impurity ions.

The values of the activation energy and resistivity of calcite ($CaCO_3$) are close to the corresponding values of quartz. A very high resistivity is observed for crystals of periclase (MgO) and mica. Thus the resistivity of muscovite mica is 10^{11} $\Omega \cdot$cm even at 500°C and the resistivity of MgO is about 10^9 $\Omega \cdot$cm at 1000°C.

§2. Formation of Space Charge and Related Phenomena

Experiments show that a current flowing through a crystal varies with time. This can be seen particularly clearly at low temperatures. Investigations of the conduction process and of the causes of the decay of a current with time were carried out by A. F. Ioffe and his colleagues and later by B. M. Gokhberg, K. D. Sinel'nikov, G. I. Skanavi, and others. These investigations showed that secondary phenomena associated with the passage of the current through a crystal are due to the formation of a space charge.

A. Decay of Current with Time. When a constant voltage is applied to a crystal at a moderate temperature, the current passing through the crystal (the charging current) decays with time. After some time the current reaches a steady-state residual value I_{res} . When the sample is disconnected from its voltage source and its electrodes are short-circuited a current appears in the opposite direction (the discharging current) and this current gradually decays to zero.

The decay of the current with time makes it difficult to investigate the electrical conductivity because this property becomes indeterminate in view of its dependence on the duration of application of the voltage.

Determination of the exact law of the decay of the current with time is not easy because of experimental difficulties in the measurement of the initial current (at t = 0), since immediately after the application of the voltage the current decreases very rapidly. Experimental results obtained by many workers show that right down to very short time intervals (shorter than 1 sec) various dielectrics obey satisfactorily the formula for the decay of the current suggested by Pierre Curie:

$$I = at^{-n}, \tag{VIII.19}$$

where a and n are constants. However, at t = 0 Eq. (VIII.19) gives a result which is physically meaningless. Moreover, Curie's formula is not obeyed during the initial period of the order of 10^{-2}–10^{-3} sec. For example, initially the current in mica obeys the law

$$I = A (t + t_0)^{-n}, \tag{VIII.20}$$

where A and n are constants, $t_0 = 2.1 \times 10^{-2}$ sec, and n = 0.87. In general, the nature of the current decay during the first moments differs somewhat from the nature of the decay after some time.

The discharging current decays with time along a curve similar to the charging current curve.

We can easily see that, in the case of inhomogeneous dielectrics, the formation of charge at interfaces redistributes the field and causes the current to decay with time. Let us consider a two-layer dielectric in which the layers have the permittivities ε_1 and ε_2, the conductivities σ_1 and σ_2, and the thicknesses d_1 and d_2. Let us assume that an internal field is applied transversely to the planes of these layers. The initial distribution of the fields E_1^0 and E_2^0 satisfies the relationship

$$\frac{E_1^0}{E_2^0} = \frac{\varepsilon_2}{\varepsilon_1}, \qquad \text{(VIII.21)}$$

but this distribution varies with time and under steady-state conditions, corresponding to the case when the currents in the two layers are equal, the distribution satisfies the law

$$\sigma_1 E_{1\text{st}} = \sigma_2 E_{2\text{st}} \qquad \text{(VIII.22)}$$

According to Eq. (VIII.22), a change of the field intensities and a redistribution of the potential takes place in the two-layer capacitor that we are considering. The variation of the field intensity with time should be accompanied by a capacitative current (displacement current). The condition for the continuity of the currents (consisting of the displacement and conduction currents) in the layers is of the form

$$\sigma_1 E_1 + \frac{\varepsilon_1}{k}\frac{dE_1}{dt} = \sigma_2 E_2 + \frac{\varepsilon_2}{k}\frac{dE_2}{dt}, \qquad \text{(VIII.23)}$$

where k is a coefficient representing the conversion of units. The solution of this equation for E_1 (assuming that $E_1 d_1 + E_2 d_2 = U$, where U is the voltage applied to the electrodes) is given by Skanavi [16] in the form

$$E_1 = \frac{(\varepsilon_2\sigma_1 - \varepsilon_1\sigma_2)\,d_2 U}{(\varepsilon_1 d_2 + \varepsilon_2 d_1)\,(\sigma_1 d_2 + \sigma_2 d_1)}\, e^{-\frac{t}{\theta}} + \frac{U\sigma_2}{\sigma_1 d_2 + \sigma_2 d_1}; \qquad \text{(VIII.24)}$$

here, θ is the time constant of the process of field redistribution:

$$\frac{\sigma_1 d_2 + \sigma_2 d_1}{\varepsilon_1 d_2 + \varepsilon_2 d_1}\, k = \frac{1}{\theta}.$$

The expression for E_2 can be written in a similar form.

Equation (VIII.24) gives the dependence of field on time. The steady state is reached at t = ∞:

$$E_{1st} = \frac{U\sigma_2}{\sigma_1 d_2 + \sigma_2 d_1} , \left.\begin{array}{c} \\ \\ \end{array}\right\}$$
$$U_{1st} = E_{1st} d_1.$$

(VIII.25)

In practice, we reach conditions approximating very closely to the steady state in a relatively short time (of the order of minutes) because the time constant is not large. The higher the conductivity of a dielectric, the lower the value of θ and the faster is the redistribution of the field.

Using Eq. (VIII.24), we can easily find the time dependence of the current

$$j = \sigma_1 E_1 + \frac{\varepsilon_1}{k} \frac{dE_1}{dt}$$

in the form

$$j = \frac{(\varepsilon_2\sigma_1 - \varepsilon_1\sigma_2)^2 \, d_1 d_2 U}{(\varepsilon_1 d_2 + \varepsilon_2 d_1)^2 \, (\sigma_1 d_2 + \sigma_2 d_1)} \, e^{-\frac{t}{\theta}} + \frac{\sigma_1\sigma_2 U}{\sigma_1 d_1 + \sigma_2 d_2} .$$

(VIII.26)

The first term in Eq. (VIII.26) is known as the absorption current. This current is due to accumulation of a free charge at the interface between the two layers and it flows as long as the accumulation process continues. The second term is the residual current which is due to the actual conductivity of a crystal. At t = ∞, j = j_{res} . At t = 0 the current density j_0 is given by the expression

$$j_0 = \frac{\varepsilon_1^2 \sigma_2 d_2 + \varepsilon_2^2 \sigma_1 \, d_1}{(\varepsilon_1 d_2 + \varepsilon_2 d_1)^2} \, U.$$

(VIII.27)

This is the highest value of the current and represents the conduction and displacement currents.

The process of discharge of our two-layer capacitor is not of great interest and can be described by the first term in Eq. (VIII.26), with its sign reversed. The dependence of the density of the total current in a two-layer dielectric on time during charging and discharging is shown graphically in Fig. 214.

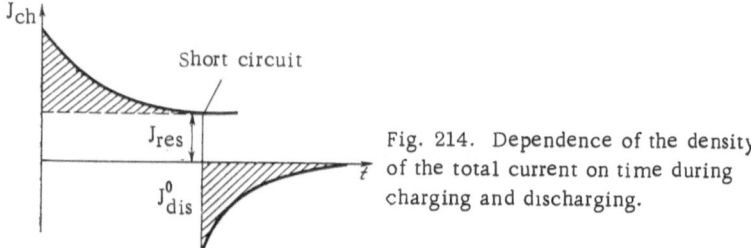

Fig. 214. Dependence of the density of the total current on time during charging and discharging.

We have already pointed out that the formation of a space charge which alters the field distribution across two layers in a capacitor is accompanied by a displacement current which can be represented by the equivalent capacitance

$$\Delta c = \frac{(\varepsilon_1 \sigma_2 - \varepsilon_2 \sigma_1)^2 \, d_1 d_2 s}{(\sigma_1 d_2 + \sigma_2 d_1)^2 \, (\varepsilon_1 d_2 + \varepsilon_2 d_1) \, k},$$ (VIII.28)

where s is the electrode area.

In some cases, the decay of the current with time is similar to the decay in a two-layer dielectric discussed in the preceding paragraphs. However, this does not mean that the decay of current with time is always due to the inhomogeneity of a dielectric. The decisive factor is the formation of a space charge. The formation of a space charge may be associated with an existing inhomogeneity as well as with an inhomogeneity which appears during the passage of a current.

In general, the accumulation of a space charge may take different forms:

1) an explicit inhomogeneity (for example, a two-layer dielectric) with the space charge accumulating at the interface;

2) the space charge may be distributed across the whole thickness of a dielectric;

3) the space charge may be formed by the accumulation of impurity ions in thin layers near the electrodes;

4) the space charge may be accumulated because poorly conducting layers are formed near one or both electrodes as a result of chemical changes (electrical "forming").

The formation of a space charge in a dielectric can be deduced from the distribution of the potential. This is frequently done by the method of probes, suggested by A. F. Ioffe. The probes are conducting strips on the surface of a dielectric whose potential is determined by connecting them to an electrometer. Examples of the distribution of the potential associated with the formation of a positive charge, a negative charge, and of charges of both signs are shown in Fig. 215.

The distributions of the potential shown in Fig. 215 can be found by solving Poisson's equation

$$\frac{\partial^2 V}{\partial x^2} = -\frac{4\pi}{\varepsilon}\rho, \qquad (VIII.29)$$

where ρ is the space charge density; x is the distance from one of the electrodes to a given point along a line of force. Assuming that ρ is constant across the volume, the solution of Eq. (VIII.29) is obtained in the form

$$V = \frac{1}{2}cx^2 + cx. \qquad (VIII.30)$$

In real dielectrics the volume distribution of ρ is not uniform. In view of this, the potential distribution curves shown in Fig. 215 are only approximate. The accumulation of a space charge (distributed nonuniformly throughout the dielectric) should be accompanied by back-diffusion of ions. The density of the current in this case is

$$j = \sigma E - D\frac{\partial \rho}{\partial x}, \qquad (VIII.31)$$

where D is the diffusion coefficient.

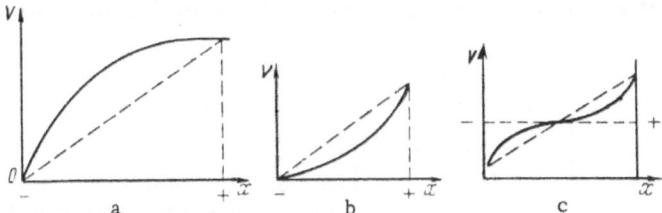

Fig. 215. Distribution of the potential within a sample in the presence of a positive (a), a negative (b), and a positive as well as a negative (c) space charge.

This equation represents the condition of continuity of the current in the case when the space charge density is not constant $(\partial\rho/\partial x \neq 0)$. When the charge density gradient is high, the field intensity is high, and the back-diffusion is fast.

The distribution of the potential in any specific case is thus governed by the distribution of the space charge. For example the distribution of the potential in a dielectric in which the space charge is concentrated in a thin layer near one of the electrodes (case 3) is of the form shown in Fig. 216a. In the case of electrical "forming" (case 4) the potential distribution is similar to that in case 3 but, because the difference between the layers is much greater, the kink in the curve is more abrupt (Fig. 216b) and the distribution is analogous to that in a two-layer dielectric (case 1).

Experimental data show that a redistribution of the potential does indeed take place during conduction in many crystals. This redistribution can be observed very clearly in calcite $(CaCO_3)$ and saltpeter $(NaNO_3)$. In these crystals charge accumulates in a thin layer near the cathode. The potential distribution is close to that shown in Fig. 216a. There are some grounds for assuming that in the cathode layer itself the space charge distribution is exponential [7]. The thickness of the layer near the cathode is estimated to be 2-3 μ.

Impurities play an important role in the accumulation of space charge. Experiments show that less space charge is accumulated in pure crystals.

Venderovich and Lapkin [3] have established that the distribution of the potential in rocksalt, fluorite, and barium oxide crystals is nearly linear if the temperature is sufficiently high (Fig. 217a). The exceptions to this rule are crystals with defects and impurities (Fig. 217b). At sufficiently high temperatures the

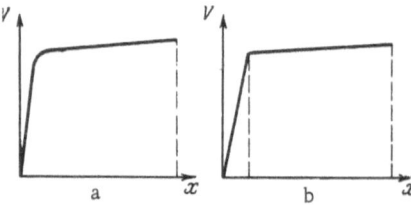

Fig. 216. Distribution of the potential within a sample when a space charge accumulates in a thin layer near an electrode (a) and after "forming" (b).

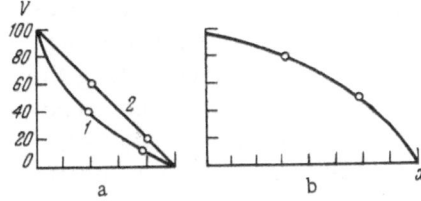

Fig. 217. Distribution of the potential in NaCl crystals: a) at temperatures of 120 and 165°C (curves 1 and 2, respectively); b) in crystals containing inhomogeneities (room temperature).

linear distribution of the potential in pure cubic crystals corresponds to a slow decay of the current with time (for example, a decay by a factor of 3-4). It should be mentioned that the kink in the dependences ln $\sigma = f(1/T)$ for alkali halide crystals is not, in general, related to the nature of the potential distribution. It is thus evident from Fig. 217a that above 165°C the distribution of the potential in NaCl is nearly linear, while a kink in the temperature dependence of the conductivity is observed at 500-700°C.

Quartz crystals exhibit a fairly slow decay of the current with time. The distribution of the potential in quartz ingots with space charges of both signs is formed during the flow of the current and these charges are distributed diffusely throughout the whole volume. The distribution is close to that shown in Fig. 215c.

Attempts have been made to relate the nature of the conductivity of a crystal, the decay of the current with time, and the distribution of the potential with the crystal symmetry. Such a correlation has no satisfactory theoretical justification and all that we can deduce from the crystal symmetry is that the electrical conductivity of all crystals (except the cubic crystals) is anisotropic. An idea of this anisotropy can be obtained from Fig. 218,

Fig. 218. Dependence $\log \sigma = f(1/T)$ for $Li_2SO_4 \cdot H_2O$ along the X (1), Y (2), and Z (3) axes [8].

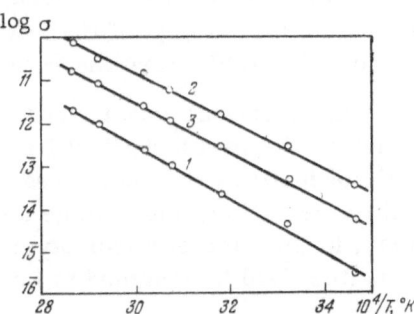

which shows the temperature dependence of the electrical conductivity of lithium sulfate monohydrate $LiSO_4 \cdot H_2O$. It is interesting
that this crystal exhibits unipolar electrical conductivity along the
Y axis (the twofold symmetry axis which coincides with the direction of the spontaneous polarization of this pyroelectric crystal).
The electrical conductivities parallel and antiparallel to the spontaneous polarization direction differ by about 10%.

To conclude our discussion of the conductivity of homogeneous dielectrics we must add that the decay of the current with time
in such materials may be divided into fast and slow processes.
The fast processes are associated with the establishment of electric polarization, which depends on the thermal motion (the thermal ionic and thermal orientational polarizations, described in
Chap. VII). The slow processes are associated with the accumulation of a space charge and electrical "forming" (in the limiting
case of chemical changes).

The decay of the current with time is accompanied by an increase in the capacitance of a condenser filled with a given dielectric because of the accumulation of a space charge. In the case
of "forming," this increase of the capacitance can be calculated
from Eq. (VIII.28). If the electrical conductivity of a "formed" layer is low, the capacitance of a sample is close to the capacitance
of this layer. This gives rise to incorrect values of the capacitance
obtained in static measurements because the results are then likely to be applied to the whole sample, while they actually refer to
the "formed" layer. In the case of a diffuse space-charge distribution the concept of an additional capacitance associated with a
particular layer loses its meaning.

B. "True" and Residual Resistance. At low
temperatures, in view of the decay of the current with time, we
must refine the concept of the resistance of a crystal since the application of a field alters the ratio of the voltage to the current.

The electrical conductivity of an inhomogeneous dielectric
depends on the parameters of its components (σ and ε). The electrical conductivity can be represented by the residual resistance,
equal to the ratio of the voltage and the residual current. In particular, in the case of a two-layer sample with the parameters σ_1
and σ_2 (the field is assumed to be normal to the interface between

the layers), the residual resistance R_{res} is given by the relationship

$$R_{res} = \frac{U}{I_{res}} = \frac{U}{i_{res}s} = \frac{\sigma_1 d_2 + \sigma_2 d_1}{\sigma_1 \sigma_2 s}, \qquad (VIII.32)$$

where s is the electrode area; d_1 and d_2 are the thicknesses of the two layers. Since in real dielectrics the current decays quite rapidly (0.1–2 min), the value of R_{res} is of practical importance.

In the case of inhomogeneous dielectrics with a more complex distribution of the components the residual resistance cannot be calculated easily from the characteristics of the components (such as their electrical conductivity or concentration), because the electrical conductivity depends strongly on the nature of the inhomogeneity (the size, shape, orientation of the particles, etc.). In general, the problem of the electrical conductivity of an inhomogeneous dielectric is equivalent to the problem of determination of the permittivity of an inhomogeneous material from the specific characteristics of its components (permittivity and concentration). The results of many investigations of inhomogeneous dielectrics are given in [9].

In the case of homogeneous dielectrics which contain defects originally or in which defects are formed during the flow of a current, we can use the concept of the "true" or "intrinsic" resistance, i.e., the resistance which represents objectively the original crystal. The basic difficulty is to determine experimentally the "true" resistance. The current is initially the sum of the capacitative (displacement) and conduction currents and, therefore, the "true" resistance R_{tr} is not equal to the initial resistance R_{in}, defined as the quotient of the voltage and the initial current I_{in}. Only in some cases do we have $R_{in} = R_{tr}$. One such case is a two-layer dielectric, consisting of layers with different conductivities but identical permittivities. $R_{in} = R_{tr}$ also applies to a two-layer dielectric in which one of the layers has a thickness and conductivity much higher than the other layer. Such conditions are satisfied, for example, by crystals of calcite and saltpeter (the space charge is concentrated in a thin layer).

Measurements of the initial current are difficult because of the very rapid decay of the current during the first moments after the application of the voltage. This makes it difficult to determine R_{tr} even when it is equal to R_{in}. Ioffe [12] has shown that the "true" resistance can be determined by measuring the potential distribution as a function of time. The accumulation of a space charge in a dielectric produces a field which is opposite in direction to the external field and, therefore, it reduces the current. The potential drop due to the space-charge field is called by Ioffe the polarization emf P.

On this basis, the current in a solid dielectric can be determined as follows:

$$I = \frac{U - P}{R_{tr}}.$$ (VIII.33)

The value of P increases with time and the current decreases. Initially P = 0. Two methods are usually employed to measure P. One of them is based on the compensation (at some moment t) of the polarization emf P by a suitably selected voltage U_1 which is exactly equal to the polarization emf P_t at that moment. The voltage U_1 thus corresponds to zero current through the dielectric and gives the value of the polarization emf P. To avoid errors the current must be measured immediately after the potential drop before redistribution of the space charge.

The second method of determination of the polarization emf is based on the measurements of the current before and immediately after a change in the external applied voltage. A sudden change of the applied voltage produces a sudden change in the current, but it does not alter the value of the polarization emf. In this case,

$$\frac{I_1}{I_2} = \frac{U_1 - P}{U_2 - P},$$ (VIII.34)

where I_1, U_1, and I_2, U_2 are, respectively, the current and voltage before and after the sudden change.

The value of P can be calculated quite easily from Eq. (VIII.34). Measurements of this kind show that P can have very large values. Therefore, the phenomenon of the formation of an opposing space-charge field is frequently called the high-voltage polarization. Thus, for quartz the polarization emf in some experiments may reach 7000 V.

In the first approximation (in those cases when $R_{tr} = R_{in}$) the polarization emf is proportional to the applied voltage U. In fact, it follows from the self-evident relationships

$$I_{in} = \frac{U}{R_{in}}$$

and

$$I_{res} = \frac{U - P_{max}}{R_{tr}}$$

that

$$P_{max} = \frac{I_{in} - I_{res}}{I_{in}} U. \qquad \text{(VIII.35)}$$

It is evident from Eq. (VII.32) that the initial current in a two-layer dielectric with $\sigma_1 \ll \sigma_2$ and $d_1 \ll d_2$ is given by

$$I_{in} = \frac{\sigma_2}{d_2} Us. \qquad \text{(VIII.36)}$$

Thus, we obtain from Eqs. (VIII.35), (VIII.32), and (VIII.36):

$$P_{max} = \frac{\sigma_2 d_1}{\sigma_1 d_2 + \sigma_2 d_1} U \equiv \alpha U, \qquad \text{(VIII.37)}$$

i.e., P_{max} is directly proportional to U. Thus, under these conditions the decaying current obeys Ohm's law. The residual current can be expressed as follows:

$$I_{res} = \frac{U - P_{max}}{R_{tr}} = \frac{U(1 - \alpha)}{R_{tr}}. \qquad \text{(VIII.38)}$$

The values of P_{max} and R_{tr} can be related to the characteristics of the two layers. This treatment can be extended from two-layer to dielectrics subjected to electrical "forming."

When the applied voltage is increased the proportionality between P_{max} and U is no longer observed. At high voltages the value of P_{max} ceases to depend on U and exhibits saturation (Fig. 219).

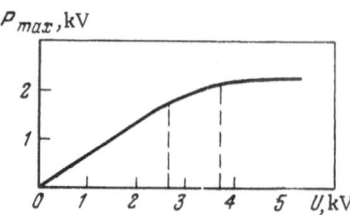

Fig. 219. Dependence $P_{max} = f(U)$ for calcite.

Knowing the polarization emf, we can easily determine the "true" resistance of a crystal using Eq. (VIII.33). Measurements show that the value of R_{tr} does not vary greatly over a wide range of voltages. For example, the "true" resistance of saltpeter crystals remains the same when the voltage is increased by a factor of 10 (from 100 to 1000 V). Further tenfold increase of the voltage (from 1000 to 10,000 V) reduces R_{tr} by a factor of 4. Investigations of rocksalt also show that the ratio P/U decreases when the field intensity is increased ($P/U = 0.17$ for $E = 10^4$ V/cm and $P/U = 0.02$ for $E = 2 \times 10^4$ V/cm).

We must mention that the accumulation of a space charge in a crystal may distort the temperature dependence of its resistance if the resistance is deduced from the residual current. The value of the resistance measured in this way may vary more strongly than the "true" resistance. When the temperature is increased we find that, first, the electrical conductivity of a dielectric increases and, secondly, the space charge is gradually dispersed so that the value of the opposing field (which reduces the current) decreases. The value of P_{max} decreases with increasing temperature and vanishes altogether at some temperature. This is because at sufficiently high temperatures ions cannot be trapped by defects for a sufficiently long time and the space charge disperses completely. For these reasons the low-temperature value of R_{res} is higher than R_{tr} but at high temperatures R_{res} and R_{tr} are equal.

C. Electrical Conductivity in Strong Fields. Electric Breakdown. We have mentioned earlier that when the polarization emf is taken into account, Ohm's law is found to be obeyed by pure crystals right up to fairly high values of the field ($\sim 10^4$ V/cm). Experiments also show that the electrical conductivity of natural crystals depends strongly on the impurities they contain. The process of electrical conduction may be accompanied by electrolytic purification, which can alter the value of the electrical conductivity. In some cases, the "true" resistance of a crystal is found to be independent of the field (up to high fields) after simple heating.

A theoretical approach shows that the current density in an ionic dielectric ceases to depend linearly on the field when the work done by the field in moving an ion becomes comparable with the energy of its thermal motion. Calculations show that, in the case of pure crystalline inorganic dielectrics, this should happen in fields of about 4×10^5 V/cm. Bearing in mind the presence of defects, we may assume that ionic crystals should obey Ohm's law right up to fields of 50-100 kV/cm. Such field intensities are close to the breakdown fields of real crystals with defects. Departure from Ohm's law in solid dielectrics was discovered by Poole in 1921 [24], who established an empirical dependence of the electrical conductivity, deduced from the residual current, on the field intensity:

$$\sigma_{res} = ae^{bE}, \qquad\qquad (VIII.39)$$

where a and b are constants.

Investigations have shown that some crystalline dielectrics exhibit an increase of the residual electrical conductivity when the field is increased, but quite often this increase does not obey Eq. (VIII.39). Better agreement between the experimental data and Eq. (VIII.39) is found in fields of moderate intensity. In weak fields the residual electrical conductivity sometimes increases with the field intensity more rapidly than predicted by Eq. (VIII.39), whereas in stronger fields the observed increase is weaker than that given by Eq. (VIII.39). The constant b decreases with increasing temperature because (due to the dispersal of the space charge) the residual electrical conductivity approaches the "true" value, which depends less strongly on the voltage. However, experiments show that the rise of the residual electrical conductivity with increasing field intensity cannot be simply attributed to the relatively less important role played by the space charge at high field intensities. An increase of the intrinsic ionic electrical conductivity may be important in strong fields.

The formula (VIII.39) is not obeyed by dielectric crystals in strong fields, particularly by crystals with a relatively low elec-

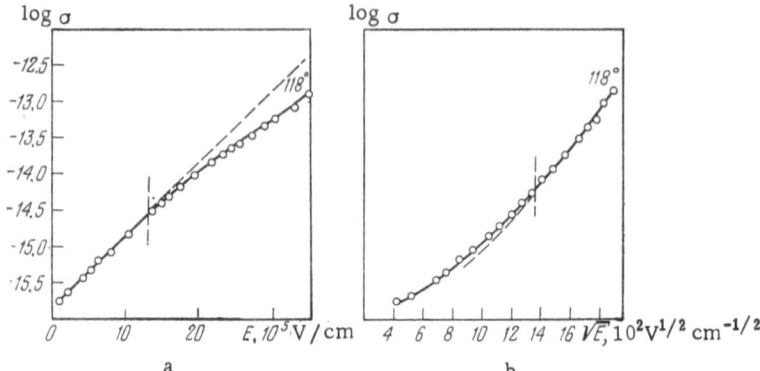

Fig. 220. Dependence of the logarithm of the electrical conductivity of mica on the field intensity E(a) and on $\sqrt{E}$ (b) [6].

trical conductivity (Fig. 220a). Frenkel [20b] has shown that in
strong fields a crystalline dielectric has an appreciable electronic
conductivity. This is because a strong field reduces the potential
barrier which an electron has to overcome in the thermal ionization process. The Frenkel formula is of the form

$$\sigma = \sigma_0 e^{\frac{1}{kT}\sqrt{\frac{e_0^2 E}{\varepsilon}}}, \qquad\qquad (VIII.40)$$

where e_0 is the electronic charge and ε is the permittivity. The
formula (VIII.40) has been confirmed experimentally for mica
(Fig. 220b), rocksalt, and other crystals. It follows that in strong
fields crystals can exhibit the electronic as well as the ionic electrical conductivity.

When the field is so strong that the work done by the field
over the mean free path of a weakly bound ion is comparable with
the energy of its thermal motion and the electronic conductivity is
superimposed on the ionic component, we may observe electric
breakdown in a crystal [5, 18, 19].

The breakdown of solid dielectrics is of two main types:
thermal and electrical. The thermal breakdown is due to the fact
that at some field intensity a dielectric is no longer in the state of
thermal equilibrium; its temperature begins to rise rapidly and
the dielectric breaks down. The departure from thermal equilibri-

um in a dielectric can also occur when the electrical conductivity increases with increasing temperature, which is typical of the ionic electrical conductivity.

In the electrical breakdown mechanism the electronic current in a dielectric crystal begins to rise rapidly and uncontrollably at some field intensity.

§3. Special Features of Electrical Conduction in Ferroelectrics

The main features of electrical conduction in ferroelectrics are the same as those of conduction in ordinary dielectrics. Ionic conduction in ferroelectrics may be intrinsic or associated with impurities or defects. Such conduction is typical of ferroelectrics with orderable structure elements.

However, electrical conduction in ferroelectrics also has a number of special features. These features are primarily associated with the presence of a domain structure and with phase transitions. Moreover, nonstoichiometry, which is practically always observed in oxygen-octrahedral ferroelectrics (it is usually a deficiency of oxygen), gives rise to electron or hole conduction. Such conduction is particularly important in surface layers of oxygen-octahedral ferroelectrics (Chap. VII). However, the nonstoichiometry of such crystals is also fairly pronounced in the interior and electron or hole conduction in these crystals is stronger than ionic conduction. Thus, we can speak of the semiconducting properties of oxygen-octahedral ferroelectrics.

A. Oxygen-Octahedral Ferroelectrics. The majority of oxygen-octahedral ferroelectrics (barium and strontium titanates, lead metaniobate, cadmium pyroniobate, lithium and potassium tantalates) are good insulators: their electrical conductivity does not exceed $10^{-12} \; \Omega^{-1} \cdot cm^{-1}$. Some oxygen-octahedral ferroelectrics have a somewhat higher electrical conductivity of about $10^{-8} \; \Omega^{-1} \cdot cm^{-1}$ (lead titanate, bismuth ferrite) but they are still insulators.

Investigations of $BaTiO_3$ single crystals show that this ferroelectric has semiconducting properties, a forbidden band 3.1 eV wide, and n-type conduction. The value of the activation energy, calculated from the slopes of the $\ln \sigma = f(1/T)$ dependences, is

1-2 eV. The fact that the activation energy is smaller than the for-
bidden-band width is usually attributed to the presence of impurities and (which is more important) a deficiency of oxygen. The
donor levels due to oxygen vacancies in the barium titanate lattice
lie at a depth of about 2.5 eV below the bottom of the conduction
band, i.e., about 0.5 eV above the top of the valence band. The activation energy of carriers in oxygen-deficient samples is 1.5-1.6
eV. Impurity levels in oxygen-octahedral ferroelectrics may also
be due to an excess of oxygen.

Calculations show that the low-temperature intrinsic electrical conductivity of real $BaTiO_3$ crystals is negligibly small compared with the conductivity associated with impurities or defects.

Numerous investigations have established that the electrical
conductivity of oxygen-octahedral ferroelectrics depends on time;
at a fixed value of the field the electrical conductivity of such ferroelectrics increases and, after a certain time, the ferroelectric
breaks down. Such phenomena are known as the aging effects.

The characteristic features of aging can be summarized as
follows. Aging is very slight (sometimes impossible to measure)
in low fields and at low temperatures; however, the electrical conductivity may increase by many orders of magnitude when the field
and temperature are increased. However, there is an upper temperature limit above which aging is not observed, irrespective of
the value of the field applied to a ferroelectric. The aging process exhibits saturation: at a fixed temperature and in a constant
field aging stops after a time interval which may range from minutes to hours or even days. An increase of the temperature or
field starts the aging process again and, once more, saturation is
reached after some time interval. Aging alters the color of the
samples and the color "front" moves from the cathode to the anode.
Aging reduces the activation energy of electrical conduction.

Heating above the critical aging temperature re-establishes
the initial properties (electrical conductivity and color) of a sample. A similar recovery can be produced by illumination with infrared radiation or by irradiation with soft x rays, as well as by
annealing in an oxygen atmosphere. Barium titanate crystals
grown by the Remeika method (with a negligible concentration of
iron) show practically no aging.

All the properties acquired by an aged ferroelectric (color, higher electrical conductivity, ways of re-establishing the initial properties, critical temperature) are fully analogous to the properties which are exhibited by a ferroelectric with a deficiency of oxygen. This allows us to associate directly the aging process with the deficiency of oxygen.

The aging mechanism can be understood taking into account the fact that the donor levels formed because of the oxygen deficiency are, to a greater or lesser extent, empty (depending on the degree of the oxygen deficiency) except when the deficiency exceeds 10^{-3}-10^{-2} %, i.e., when degeneracy takes over. Since the color front moves from the cathode to the anode and the corresponding optical absorption is associated with the presence of electrons at the donor levels, it is evident that electrons which fill the empty levels are injected by the cathode when a field is applied to it. We must reject the hypothesis that the donor levels are filled with conduction-band electrons because to produce even a slight (of the order of 1%) excess electron density in the conduction band near the cathode would require fields of the order of 10^5-10^6 V/cm, whereas aging is observed in much weaker fields. We must therefore assume that electrons are transferred from the cathode directly to the donor levels near the cathode, bypassing the conduction band by the tunnel effect. This mechanism is supported also by a strong dependence of the intensity of the aging process on the field and the concentration of the empty levels. After reaching these empty levels, some electrons are transferred by thermal fluctuations to the conduction band, where they either participate in conduction, giving rise to a higher electrical conductivity, or migrate along the direction of the field and are trapped at the empty levels located further away from the cathode (the color front moves from the cathode to the anode). Some of these trapped electrons are, in their turn, transferred by thermal motion to the conduction band where they either participate in conduction (electrical conductivity increases still further) or they migrate in the field and are trapped by the empty levels situated even further away from the cathode. This continues until a dynamic equilibrium is established between the process of injection of electrons by the cathode into the nearby empty levels, the transfer of electrons from the filled levels to the conduction band, and the trapping of conduction electrons by the empty donor levels. The second and third pro-

cesses are governed by the temperature and the third is also influenced by the concentration of the empty levels. The first process is governed by the field and the concentration of the empty
levels.

The low intensity of the aging process at low temperatures
and fields may be explained by the low probability of the tunnel effect (in the first process) and the low probability of electron transfer from the filled levels to the conduction band (the second process). The critical temperature, above which no aging is observed,
can be explained by a low probability of the trapping of the conduction electrons by the empty levels (the third process). Dynamic
equilibrium of all three processes corresponds to saturation of the
aging of the electrical conductivity.

The aging process can be suppressed by preparing ferroelectric materials in such a way that they are not deficient in oxygen. We can also introduce impurities which compensate the oxygen deficiency: for example, a very small amount of iron in barium titanate compensates the oxygen deficiency. The electrical neu-
trality of the lattice, deficient in oxygen ions, is conserved by a
change in the charge state of ions surrounding an oxygen vacancy;
such a change is exhibited particularly by titanium ions.

The electrical conductivity is increased by the presence of
the following impurities: silver, bismuth, cobalt, gallium, man-
ganese, molybdenum, antimony, niobium, tantalum, zinc, iron (in
amounts up to a few percent), chromium, and vanadium. The electrical conductivity is increased particularly strongly by the introduction of rare-earth impurities (lanthanum, cerium, and samari-
um) in optical amounts of 0.1-0.3 mole %. These impurities alter
the color of barium titanate crystals to blue, due to transitions of
some quadruply charged titanium ions to the triply charged state.
Investigations of the luminescence spectra of barium titanate crystals containing rare earths show that rare-earth ions replace titanium ions (this is true also of strontium titanate crystals) and
not barium ions, as might be expected from a comparison of the
corresponding ionic radii. The depth of the donor levels due to
rare-earth impurities is about 2.5 eV, i.e., it is the same as that
of the levels due to the oxygen deficiency.

Sometimes the color front in biased $BaTiO_3$ single crystals
moves from the anode to the cathode. This process can be ex-

plained by the presence of cation defects in the structure. In such crystals the field drives electrons to the anode and holes are captured by cation vacancies (p-type conduction); this process colors a crystal. Some crystals exhibit different types of conduction at different temperatures. Usually, aged samples of oxygen–octahedral ferroelectrics exhibit n-type conduction in a wide range of temperature. Un-aged samples of the same ferroelectrics usually exhibit p-type conduction below 250–280°C and n-type conduction above these temperatures. Intrinsic ionic conduction may play the dominant role in some samples at high temperatures.

Measurements of the Hall effect in $BaTiO_3$ crystals with an oxygen deficiency show that the electron mobility in the conduction band is about $0.1 \ cm^2 \cdot V^{-1} \cdot sec^{-1}$.

B. Ferroelectrics with Orderable Structure Elements. We have mentioned earlier that ferroelectrics with orderable structure elements usually exhibit ionic conduction.

TABLE 45. Some Electrical Conduction Parameters of Ferroelectrics with Orderable Structure Elements

Ferroelectric	Activation energy of ionic electrical conduction, eV	Forbidden band width (optical measurements), eV
Guanidine aluminum sulfate hexahydrate, Z-cut	1.25 (14—87° C)	5.5
Triglycine sulfate X-cut Y-cut Z-cut	0.6 in ferroelectric phase 0.4 in paraelectric phase	6.0
Rochelle salt X-cut	0.9 in ferroelectric phase ~ 2.0 in paraelectric phase	Not determined (> 5)
Sodium nitrite Y-cut	~ 1 in ferroelectric phase ~ 1.5 in paraelectric phase	— 5.5
Potassium and ammonium dihydrogen phosphates KH_2PO_4 $NH_4H_2PO_4$	Investigated only in paraelectric phases: ~ 0.8-1.0	Not determined (> 5.0)

Some data on the electrical conductivity of ferroelectrics of this type are given in Table 45.

The high resistivity of ferroelectrics with orderable struc-ture elements ($\sim 10^{13}$–10^{15} $\Omega \cdot$ cm) prevents us, in practice, from checking whether they obey Faraday's law. This law cannot be verified either at higher temperatures (at which the electrical con-ductivity of these substances is somewhat higher) because such ferroelectrics usually have a low thermal stability.

The wide forbidden band (> 5.0 eV) and the absence of opti-cal absorption corresponding to electronic conduction show that the conduction process in these ferroelectrics is ionic. Differ-ences between the electrical conductivity of different samples of the same ferroelectric show that ionic conduction is due to im-purities. This applies, for example, to triglycine selenate and mixed crystals of triglycine selenate with triglycine sulfate.

The data on triglycine sulfate are more precise. These crys-tals are usually grown from solutions containing an excess (above the stoichiometric ratio) of sulfuric acid. The electrical conduc-tivity of such crystals depends strongly on the number of excess sulfate ions. Consequently, we may assume that the extrinsic (impurity) ionic conductivity of triglycine sulfate crystals, grown by this method, is — to a great extent — governed by the presence of sulfate impurity ions. The extrinsic conductivity of other fer-roelectrics listed in Table 45 is most likely to be due to the pres-ence of accidental impurities. This is confirmed by the observa-tion that multiple recrystallization reduces their electrical con-ductivity quite strongly.

Impurities are distributed nonuniformly in these crystals. Studies carried out on crystals of guanidine aluminum sulfate show that impurities are usually concentrated at the boundaries of growth pyramids, where the electrical conductivity is higher.

Since the majority of crystals with orderable structure ele-ments are organic compounds, it is most likely that impurities produce Frenkel effects in these compounds: impurity ions are located at interstices of the host lattice. In some organic ferro-electrics with hydrogen bonds we must take into account the pos-sibility of intrinsic conduction. In such cases, the conduction pro-cess is due to protons of the hydrogen bonds (the vacancy mecha-nism). Thus, for example, lithium sulfate, which is a pyroelec-

tric with hydrogen bonds, exhibits an intrinsic electrical conduction (Fig. 218). Crystals of this compound with the same orientation have the same values of the electrical conductivity. It has
been shown experimentally that the intrinsic conduction process
in these crystals is due to protons of the hydrogen bonds. Intrinsic electrical conduction in ferroelectrics such as sodium nitrite
or potassium iodate is most probably due to alkali ions.

C. Special Features Associated with Ferroelectric Phase Transitions. The temperature dependence of the electrical conductivity of ferroelectrics can be described (in most cases) by the two-term formula (VIII.14). The
most interesting and important is the observation that the dependence $\ln \sigma = f(1/T)$ has a kink which corresponds to a ferroelectric phase-transition temperature. This means that a structural
change near a phase transition and the appearance (disappearance)
of spontaneous polarization are accompanied by a change in the
activation energy of carriers. Some ferroelectrics exhibit not only a change in the activation energy but also a conductivity discontinuity in the phase-transition region.

This can be seen in Fig. 221a, which shows the temperature
dependence of the electrical conductivity of $BaTiO_3$ crystals. A

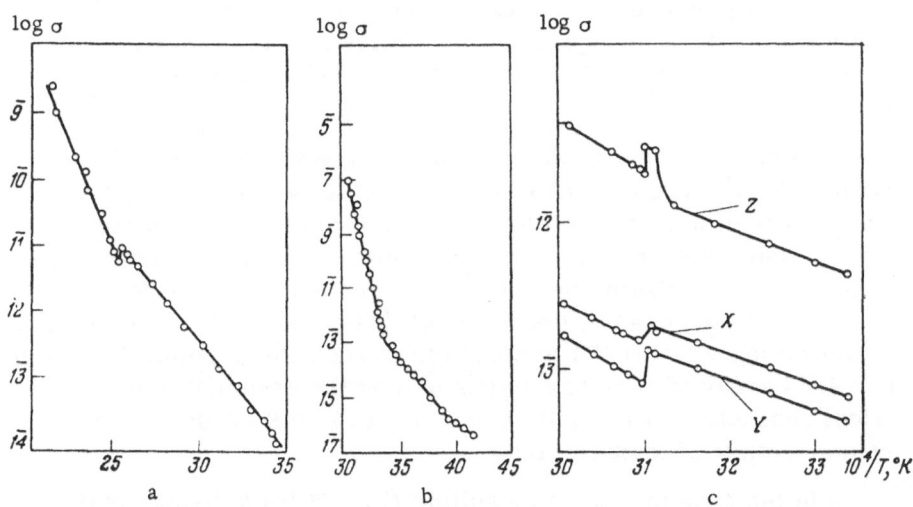

Fig. 221. Dependences $\log \sigma = f(1/T)$ for crystals of $BaTiO_3$ (a), Rochelle salt (b),
and triglycine sulfate (C).

kink near the phase transition point can be seen quite clearly. The activation energy of electrical conduction in the ferroelectric phase is less than the activation energy in the paraelectric state. A kink in the log $\sigma = f(1/T)$ dependence of $BaTiO_3$ is also observed in the region of the transition from the tetragonal to the orthorhombic modification. Similar kinks have been reported for layered crystals of bismuth titanate ($Bi_4Ti_3O_{12}$), mixed crystals of barium titanate and hafnate, and for other crystals.

The conclusion that a kink in the dependence ln $\sigma = f(1/T)$ corresponds to a phase transition temperature can also be proved indirectly. Thus, for example, a study of the temperature dependence of the permittivity of barium titanate crystals (carried out in a thermostat in which the temperature was held constant to within $\pm 0.1°C$ and varied in steps of about 0.4°C) has indicated a thermal hysteresis of the Curie point (120°C) amounting to 4-6°C. A study of the electrical conductivity of the same crystals under the same conditions has indicated a thermal hysteresis of the electrical conductivity anomaly identical (within the limits of the experimental error in measurement of the temperature) with the thermal hysteresis of the Curie point.

The changes in the activation energy observed near the phase transition points of oxygen–octahedral ferroelectrics are evidently associated not only with a change in the forbidden band width but also with a change in the electronic component of the permittivity (i.e., with a change in the refractive index n) since this component determines the depth of impurity levels in the forbidden band.

Ferroelectrics, in which a phase transition involves the ordering of structure elements, also exhibit a change in the activation energy of carriers at the transition temperature and, sometimes, also a discontinuity of the absolute value of the electrical conductivity σ. Examples of such changes are shown in Figs.221b and 221c. In the case of Rochelle salt (Fig. 221b) a kink in the dependence log $\sigma = f(1/T)$ is observed at each Curie point. The activation energy of carriers, calculated on the assumption that electrical conduction in Rochelle salt is ionic, is 0.9 eV in the ferroelectric range of temperatures.

In the case of triglycine sulfate (Fig. 221c) a discontinuity of the electrical conductivity at the Curie point is clearly evident for all three crystallographic axes X, Y, and Z. It is interesting that this crystal shows a weak thermal hysteresis (~ 0.2°C) of the phase transition point, observed in measurements of the electrical conductivity. This hysteresis and the discontinuity in the electrical conductivity σ indicate that triglycine sulfate, which undergoes a phase transition of the second kind, exhibits some (although not very pronounced) features typical of ferroelectrics which undergo a phase transition of the first kind.

Assuming that electrical conduction in triglycine sulfate is ionic, we find that the activation energy of carriers is 0.4 eV in the ferroelectric region and 0.6 eV in the paraelectric region.

Although experimental data on the relationship between changes in the structure and spontaneous polarization, on the one hand, and changes in the electrical conductivity, on the other, are available, the nature of this relationship is not yet understood.

D. Absorption Phenomena Associated with the Presence of a Domain Structure. In §2 of the present chapter we have discussed, in particular, the phenomenon of dielectric absorption, which is associated with the inhomogeneity of some dielectrics and the formation of a space charge. Dielectric absorption can be observed also in ferroelectric crystals. However, in the case of ferroelectrics, slow processes of domain polarization (reorientation), which also give rise to a decay of the current with time, may be superimposed on the usual polarization and conduction processes. Moreover, in those cases when the formation of a space charge, motion of impurities, electrical "forming," etc. are unimportant, the decay of the current with time may be entirely due to the reorientation of domains. If we assume that the decay of the current with time is due to the reorientation of domains, the mechanism of this decay can be described as follows. Since the reorientation of domains is a relatively slow process (particularly in weak fields), the application of a field to a ferroelectric produces a current in a circuit containing a battery and a ferroelectric capacitor. This current is necessary for maintaining a constant field intensity across the capacitor and, in the final analysis, it is solely due to the potential difference provided by the voltage source. In the absence of domain reorientation (and other processes which might give rise to a slow decay of the current) the current falls instantaneously to its final

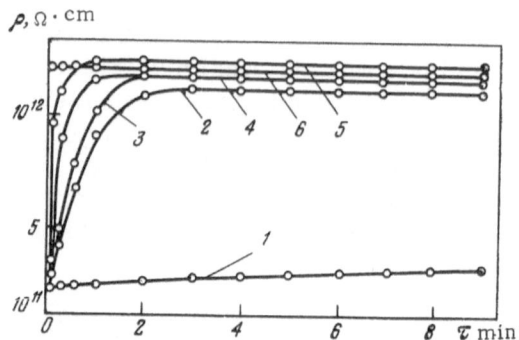

Fig. 222. Dependence of the resistivity of a triglycine sulfate crystal (Y-cut) on time for different field intensities (V/cm): 1) 10; 2) 200; 3)
400; 4) 800; 5) 1200; 6) 2400; t = 20°C.

value, whereas in the presence of domain reorientation the current decays to its final value throughout the duration of the reorientation process.

A slow decay of the current with time is observed in BaTiO$_3$,
triglycine sulfate, guanidine aluminum sulfate, and Rochelle salt,
but only in the ferroelectric range of temperatures. This decay
is negligible in weak fields (weaker than the coercive field), it is
appreciable and fairly slow (2-3 min) in fields close to the coercive
field, and is considerable as well as fast (the decay is completed
in a few seconds) in fields higher than the saturation polarization
field. This is illustrated in Fig. 222, which shows the decay of the
current with time in a triglycine sulfate crystal as a time dependence of the resistance (this is not the "true" resistance, which
can be estimated from the current in weak fields represented by
curve 1 in Fig. 222). Similar dependences are obtained also for
BaTiO$_3$, Rochelle salt, and guanidine aluminum sulfate. It has been
found that the decay of the current with time in these crystals is
not observed for nonferroelectric cuts and for ferroelectric cuts in
those cases when a crystal had been subjected to a strong pre-
liminary polarization. All these observations show that the slow
decay of the current with time in some ferroelectrics is practically entirely due to the processes of slow reorientation of domains.
Consequently, the slow decay of the current with time in ferroelectrics associated with domain polarization may be called the

ferroelectric absorption. It must be stressed that this phenomenon is not associated with the formation of a space charge in a crystal.

§4. Dielectric Losses (General Discussion)

An alternating electric field heats any real dielectric. That part of the energy of the alternating electric field which is transformed into heat is known as dielectric losses. The total dielectric losses (or simply the losses) consist of conduction losses, corresponding to a constant voltage, and losses due to the active component of the displacement current.

A. Establishment of Polarization. Dielectric losses are associated with the process of the establishment of polarization. As shown in Chap. VII, polarization is not established instantaneously in a dielectric: a finite time interval passes between the moment of application of a field and the moment of establishment of a steady state.

The displacement polarization is established rapidly. The electronic displacement polarization appears in a time comparable with the period of optical-frequency oscillations (10^{-14}-10^{-15} sec). The ionic displacement polarization is established in a time comparable with the period of natural oscillations of ions about their equilibrium positions (10^{-12}-10^{-13} sec). The dispersion of the permittivity ε, associated with the ionic displacement polarization, is observed at long infrared wavelengths (10-100 μ). Thus, the displacement polarization is established in a time which is very short compared with the period of fields used in electrical and radio engineering (including microwave frequencies).

It follows that the establishment of the displacement polarization at frequencies used in electrical and radio engineering does not give rise to any appreciable energy dissipation. The dielectric losses in crystals which exhibit only the displacement polarization are very low (an estimate of these losses will be given later).

The situation is somewhat more complicated in the case of dielectrics exhibiting the relaxational polarization, particularly in the case of dielectrics exhibiting polarization associated with thermal motion (thermal orientational and thermal ionic polarization). Numerous dielectrics, exhibiting these types of polarization, have fairly high losses at radio frequencies because the time for the establishment of polarization is now comparable with the period of an external electric field.

We shall begin our discussion of the processes of establishment of the thermal orientational polarization in solid dielectrics by considering the results obtained by Debye in 1929, who developed a theory of dielectric relaxation in polar liquids and in solutions of polar molecules in nonpolar solvents. Considering polar molecules as spheres rotating in a viscous medium, Debye obtained an expression for the relaxation time τ of dipoles

$$\tau = \frac{\xi}{2kT} = \frac{4\pi\eta r^3}{kT}, \qquad (VIII.41)$$

where ξ is the coefficient of friction opposing rotational motion of a sphere in a viscous medium, given by the Stokes law ($\xi = 8\pi\eta r^3$); η is the viscosity; r is the radius of a sphere (molecule). According to this theory, the frequency dependence of the orientational polarization α_{or} is of the form

$$\alpha_{or} = \frac{\alpha_{0or}}{1 + \iota\omega\tau}, \qquad (VIII.42)$$

where α_{0or} is the static orientational polarization [see Eq. (VII.42)].

The Debye approximation is fairly inaccurate. It is difficult to imagine molecules as solid spheres moving in a viscous medium. The concept of viscosity is hardly applicable to the "friction" between molecules. Only in the case of large polar molecules, which can exhibit gradual reorientation, is the motion during polarization reversal similar to the motion in a viscous medium.

The Debye value of the relaxation time of dipoles τ is, in general, not equal to the time for establishment of the polarization θ. The value of θ can be found bearing in mind that the field intensity acting on a dipole varies during polarization. When the Lorentz calculation of the effective field is valid [see Eq. (VII.14)], i.e., when the Clausius−Mosotti equation applies, the time θ is found to be related to the time τ by the expression (which will be discussed later)

$$\theta = \frac{\varepsilon_s + 2}{\varepsilon_\infty + 2}\, \tau, \qquad (VIII.43)$$

where ε_s is the static permittivity and ε_∞ is the permittivity due to the displacement polarization (including the elastic dipole displacement polarization). However, it must be mentioned that the Clausius–Mosotti formula is, in general, inapplicable to polar liquids. This means that Eq. (VIII.43) is of limited validity and can be used only to describe the establishment of polarization in solutions with a low concentration of polar molecules or in low-density polar gases.

The treatment of the motion of polar molecules as the motion in a viscous medium is not the only possible approach. The process of reorientation of dipoles can also be treated as a transition from one energy state to another (accompanied by a change of the direction of a dipole), the two states being separated by a potential barrier U. We can show [16] that by substituting a definite potential barrier U for the viscosity, we obtain the same expressions for the frequency dependence of α_{or} .

The Debye relationships can be used to analyze the relaxation of dipoles in nonferroelectric solids. In the simplest case we may assume that each molecule in a solid has a constant electric moment μ, which can be oriented only along two directions: parallel to a field E or antiparallel to this field. Analytically this problem is equivalent to transitions of weakly bound ions, considered in Chap. VII (§3), except that, in the present case, out of a total number of n_0 dipoles, half of them ($n_0/2$ and not $n_0/6$) attempt to "jump" the potential barrier (i.e., to reverse their direction). Consequently, Eq. (VII.47) is now written as follows:

$$\Delta n = \frac{n_0 e \delta}{4kT} (1 - e^{-\frac{t}{\tau}}) E. \qquad \text{(VIII.44)}$$

The meaning of the symbols is the same as in Eq. (VII.47).

Using the expression relating the polarization at a given moment P_t and the value of Δn [see Eq. (VII.55)]

$$P_t = \Delta n e \delta,$$

we obtain

$$P_t = \frac{n_0 e^2 \delta^2}{4kT} (1 - e^{-\frac{t}{\tau}}) E. \qquad \text{(VIII.45)}$$

502   CHAPTER VIII

The value of the orientational polarizability α_{or} can be deduced from Eq. (VIII.45) using $\mu = e\delta/2$:

$$\alpha_{or} = \frac{n_0\mu^2}{kT}\left(1 - e^{-\frac{t}{\tau}}\right). \qquad (VIII.46)$$

Using the expression for the static orientational polarizability (assuming that $\mu E \ll kT$)

$$\alpha_{0or} = \frac{n_0\mu^2}{3kT}, \qquad (VIII.47)$$

we can see that solid polar dielectrics obey the relationship

$$\alpha_{or} \cong \alpha_{0or}\left(1 - e^{-\frac{t}{\tau}}\right), \qquad (VIII.48)$$

which is similar to Eq. (VIII.42) for polar liquids.

It follows from Eq. (VIII.48) that at low frequencies ($t \to \infty$) and for moderately high fields and temperatures the orientational polarizability of solid polar dielectrics is close to the orientational polarizability of polar liquids. When the period of an alternating field t is close to the relaxation time, the orientational polarizability is about $2/3$ of the static value α_{0or}. At very high frequencies ($t \ll \tau$) the contribution of the orientational polarizability to the total effect tends to zero.

Allowance for a change in the effective field F during the establishment of polarization gives an expression for the contribution of the reorientation of dipoles to the permittivity, derived in Chap. VII [Eq. (VII.47)].

Special features of the establishment of the thermal ionic polarization are discussed in detail in Chap. VII (§3). The main result given in that chapter [Eq. (VII.53)] can be written in the form

$$P_t = \frac{n_0 e^2 \delta^2}{12kT}\left(1 - e^{-\frac{t}{\tau}}\right) E, \qquad (VIII.49)$$

where τ is given by the relationship

$$\tau = \frac{e^{\frac{U}{kT}}}{2\nu}. \qquad (VIII.50)$$

Equation (VIII.49) is identical with Eq. (VIII.45) to within a constant factor. This means that the establishment of the thermal orientational polarization (in solid and liquid dielectrics) and the establishment of the thermal ionic polarization are qualitatively equivalent. Numerical values of the time for the establishment of the polarization (or the relaxation time, which is – in general – different from the time of establishment of the polarization) depend on the nature of a dielectric and differ from one dielectric to another (this will be discussed later).

We note that Eq. (VIII.49) is derived on the assumption that the effective field remains constant and equal to the average macroscopic field during the polarization process and that the change in the height of the potential barrier (ΔU) is small compared with the energy of thermal motion kT. Allowance for the change in the effective field during polarization of a crystal (assuming that the Clausius–Mosotti equation applies to a given dielectric) for the displacement polarization makes it possible to determine the relationship between the time for the establishment of the polarization θ and the relaxation time of ions τ:

$$\theta = \tau \frac{\varepsilon_0 + 2}{\varepsilon_\infty + 2} = \frac{e^{\frac{U}{kT}}}{2\nu} \frac{\varepsilon_s + 2}{\varepsilon_\infty + 2}. \tag{VIII.51}$$

The above expression is identical with the analogous expression (VIII.43) for polar liquids.

For the majority of crystals exhibiting the thermal ionic polarization the factor $(\varepsilon_c + 2)/(\varepsilon_\infty + 2)$ is close to unity. For some crystals and polar liquids this factor can be larger but the Clausius–Mosotti formula is inapplicable to such dielectrics and the field acting on each molecule does not differ greatly from the average macroscopic field. Therefore, Eq. (VIII.50) should give the correct order of magnitude of the relaxation time of polar liquids.

Taking into account the change in the effective field F during thermal ionic polarization, we find that the contribution of this polarization can be obtained from the relationship

$$P_t = \frac{n_0 e^2 \delta^2}{12kT} \frac{\varepsilon_s + 2}{3} (1 - e^{-\frac{t}{\theta}}) E, \tag{VIII.52}$$

which can be transformed, taking into account the relationship between E and F

$$\frac{\varepsilon_s + 2}{3} E = F, \qquad (VIII.53)$$

to the form

$$P_l = \frac{n_0 e^2 \delta^2}{12kT} \left(1 - e^{-\frac{t}{\theta}}\right) F. \qquad (VIII.54)$$

As expected, Eq. (VIII.54) is identical with Eq. (VIII.49) because the difference between the effective and average fields is ignored in the derivation of the second of these equations and the effective field is assumed to be constant. However, allowance for the change in the effective field gives rise to a change in the macroscopic polarizability (due to the thermal ionic polarization) by a factor of $(\varepsilon_s + 2)/3$, as indicated by Eq. (VIII.52). Thus, the greater the difference between ε_s and unity, the more important is the allowance for the change of the effective field in the determination of the macroscopic polarizability.

It follows from Eqs. (VIII.43) and (VIII.51) that the time for the establishment of polarization (or the time constant of the process of the establishment of polarization) is always somewhat larger than the relaxation time τ. This can be understood taking into account the fact that the field acting on an ion (or a dipole) increases with time up to its steady-state value. Thus, the final polarization, governed by the effective field, is stronger than the polarization which would be obtained in a constant effective field. Naturally, the stronger polarization requires a longer time for its establishment.

The activation energy of weakly bound ions is of the order of 10^{-12} deg. The frequency of their transitions ν is about 10^{12} sec^{-1} (lower than the frequency of natural oscillations of bound ions). It follows that the relaxation time τ at 300°K, estimated using Eq. (VIII.50), should be about 3.5×10^{-2} sec. At 600°K the relaxation time should be $\tau \approx 3.1 \times 10^{-7}$ sec. These values of the relaxation time are comparable with the periods of alternating fields used in electrical and radio engineering. This means that dielectric losses in alternating fields may be considerable for materials exhibiting the thermal ionic polarization (or the thermal orientational polarization).

The polarization of a dielectric exhibiting the displacement and thermal ionic polarization mechanisms can be found from the relationship [Eqs. (VII.15) and (VIII.52)]

$$P_t = n_0\eta F + \frac{n_0 e^2 \delta^2}{12kT} \frac{\varepsilon_s + 2}{3}(1 - e^{-\frac{t}{\theta}}) E. \qquad (VIII.55)$$

The effective field F increases during the establishment of the relaxational (thermal ionic) polarization; consequently, the electric moment per unit volume due to the displacement polarization $n_0\eta F$ will also increase, in spite of the fact that η is constant. When the Lorentz formula is valid $(F = E + 4\pi/3P)$ we obtain

$$P_t = \frac{n_0\eta E}{1 - \frac{4\pi n_0\eta}{3}} + \frac{n_0 e^2 \delta^2}{12kT\left(1 - \frac{4\pi n_0\eta}{3}\right)} \cdot \frac{\varepsilon_s + 2}{3}(1 - e^{-\frac{t}{\theta}}) E. \quad (VIII.56)$$

However,

$$1 - \frac{4\pi n_0\eta}{3} = 1 - \frac{\varepsilon_\infty - 1}{\varepsilon_\infty + 2} = \frac{3}{\varepsilon_\infty + 2}$$

and, therefore [see Eq. (VII.19)],

$$n_0\eta = \frac{3}{4\pi} \frac{\varepsilon_\infty - 1}{\varepsilon_\infty + 2}. \qquad (VIII.57)$$

It follows that

$$P_t = \left[\frac{\varepsilon_\infty - 1}{4\pi} + \frac{n_0 e^2 \delta^2}{12kT} \cdot \frac{(\varepsilon_\infty + 2)(\varepsilon_s + 2)}{9}(1 - e^{-\frac{t}{\theta}})\right] E. \qquad (VIII.58)$$

It is evident from this expression that the difference between the effective and average macroscopic fields affects not only the relaxation time but the value of the electric moment per unit volume at any given moment. Comparison of Eqs. (VIII.52) and (VIII.58) shows that allowance for the displacement polarization increases the thermal ionic polarizability by a factor of $(\varepsilon_\infty + 2)/3$.

We have mentioned earlier that dielectric losses in crystals are associated with the slow processes of establishment of electric polarization. However, the slow establishment of the polarization does not necessarily imply that there are dielectric losses. These losses are due to the fact that the processes associated

with the slow establishment of the polarization are accompanied
by absorption currents. We have mentioned in §2 of the present
chapter that the processes of establishment of the thermal ionic
and thermal orientational polarization are associated with absorption currents. These processes are regarded as slow, in spite of
the fact that they are completed very rapidly compared with the
processes associated with the formation of a space charge, interfacial polarization, electrical "forming", etc., which are also associated with absorption currents.

The absorption current density can be found by differentiating the electric moment per unit volume (the polarization) with
respect to time. The application of Eqs. (VIII.45) and (VIII.46) to
the thermal orientational polarization in a solid dielectric (ignoring the difference between the effective and average fields)
gives the following expression for the absorption current density:

$$j_{abs} = \frac{n_0 \mu^2}{kT\tau} e^{-\frac{t}{\tau}} E. \tag{VIII.59}$$

In the case of the thermal ionic polarization (taking into account
the difference between the effective and average fields) the absorption current density is found by differentiating Eq. (VIII.58):

$$j_{abs} = \frac{n_0 e^2 \delta^2 (\varepsilon_s + 2)(\varepsilon_\infty + 2)}{9 \cdot 12 kT\theta} e^{-\frac{t}{\theta}} E. \tag{VIII.60}$$

These two expressions show that the absorption current should
depend exponentially on time. However, we have demonstrated earlier [see Eqs. (VIII.19) and (VIII.20)] that the absorption current
during short time intervals varies in accordance with the power
law $I = At^{-n}$ or $I = A(t - t_0)^{-n}$. This disagreement is due to the fact
that the absorption current need not be associated solely with the
polarization. Moreover, we must take into account the fact that
only in rare cases can a dielectric be represented by a single value
of the potential barrier for all its weakly bound ions (or dipoles).
In general, a given dielectric has a range of values of such barriers and this gives rise to a range of time constants of the current decay process. Allowance for this point makes it possible to
achieve agreement between the theory and the experimental data.

B. Fundamentals of the Theory of Dielectric
Losses. The dependence of the absorption current on time

[Eqs. (VIII.59), (VIII.60)] can be represented, in general, by

$$j = ge^{-\frac{t}{\theta}} E.$$

A sudden increase of the field at a moment t by an amount ΔE produces the following rise in current:

$$\Delta j = ge^{-\frac{t-t_0}{\theta}} \Delta E. \qquad (VIII.61)$$

Considering short intervals of time and going to the limit [$\Delta E = dE\,(d\tau/d\tau)$], we can use the principle of the superposition of the currents * to represent the absorption current due to all changes in the field (beginning from t $= -\infty$) to a moment t, by the following integral:

$$j = g \int\limits_{-\infty}^{\infty} \frac{dE}{d\tau} e^{-\frac{t-\tau}{\theta}} d\tau. \qquad (VIII.62)$$

In the simplest case, when the field is sinusoidal, i.e.,

$$E = E_m e^{i\omega t},$$

integration of Eq. (VIII.62) gives the expression

$$j = \frac{\omega^2\theta^2 g}{1 + \omega^2\theta^2} e^{i\omega t} E_m + i \frac{\omega\theta g}{1 + \omega^2\theta^2} e^{i\omega t} E_m. \qquad (VIII.63)$$

Thus, the absorption current in the case of an alternating voltage consists of two components. One component, represented by the first term in Eq. (VIII.63), is in phase with the voltage and is known as the active component. The other component, described by the second term, is the current which leads the voltage by a phase angle $\pi/2$ and is the reactive (capacitative) component of the current. The amplitudes of the active and reactive components of the current are

$$j_{am} = \frac{\omega^2\theta^2 g}{1 + \omega^2\theta^2} E_m; \quad j_{rm} = \frac{\omega\theta g}{1 + \omega^2\theta^2} E_m. \qquad (VIII.64)$$

---

* The principle of superposition of the currents states that the total current passing at a given moment through a dielectric after a series of different voltage changes is equal to the sum of the current due to each of the preceding voltage changes.

This discussion yields an important conclusion: if the polarization in a dielectric is established slowly, energy losses appear in alternating fields even in the absence of dc conductivity.

Figure 223 gives the vector diagram of the density of currents j_a and j_r (corresponding to the slowly established polarization) and the field intensity E in a dielectric. The angle φ represents the phase shift between the total current and the voltage. The angle δ, which is the difference between $\pi/2$ and φ, is known as the dielectric loss angle: it is evident that $\tan \delta = j_{am}/j_{rm}$.

Since, under alternating voltage conditions, a steady-state charge distribution (and a steady-state value of the polarization) cannot be established, the permittivity measured at an alternating field is lower than the permittivity measured using dc. The decrease in the permittivity ε of a dielectric in alternating fields can be due to causes other than a slowly established polarization (for example, it may be due to the fact that a space charge, formed in constant fields, cannot accumulate in alternating fields because of insufficient time). However, we shall consider here only those changes in ε which are associated with a slowly established polarization.

We shall calculate first the capacitative (reactive) component of the permittivity. Using the well-known relationship between the capacitance C_r, the amplitude of the reactive current I_{rm}, and the voltage amplitude U_m

$$I_{rm} = \omega C_r U_m,$$

we obtain the following expression for the density of the reactive

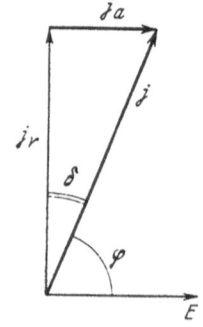

Fig. 223. Vector diagram of the current densities and field intensities in a dielectric.

current density j_{rm} :

$$j_{rm} = \frac{\omega \varepsilon_r E_m}{4\pi}. \qquad \text{(VIII.65)}$$

Taking into account only the current density j_r due to a slowly established polarization, defined by Eq. (VIII.64), and equating the right-hand sides of Eqs. (VIII.64) and (VIII.65), we obtain

$$\varepsilon_r = \frac{4\pi\theta g}{1 + \omega^2\theta^2}. \qquad \text{(VIII.66)}$$

A similar calculation of j_r can be carried out also for the displacement polarization but, in this case, the permittivity ε is independent of the frequency throughout the whole radio-frequency range because of the smallness of θ ($\sim 10^{-13}$ sec). When the voltage is constant ($\omega = 0$) we obtain from Eq. (VIII.66)

$$\varepsilon_r = 4\pi\theta g. \qquad \text{(VIII.67)}$$

It follows from Eq. (VIII.66) that, at low frequencies ($\omega\theta \ll 1$), the value of ε_r depends weakly on the frequency. In the frequency range $\theta \approx 1/\omega$ the value of ε_r decreases and at high frequencies ε_r varies inversely proportionally to ω^2.

Considering the density of the active current in Eq. (VIII.64) as being due to the "active capacitance" of a dielectric, with a permittivity ε_a, i.e.,

$$j_{am} = \frac{\omega \varepsilon_a E_m}{4\pi}, \qquad \text{(VIII.68)}$$

we obtain an expression for ε_a by comparing j_{am} given by Eqs. (VIII.64) and (VIII.68):

$$\varepsilon_a = \frac{4\pi\omega\theta^2 g}{1 + \omega^2\theta^2}. \qquad \text{(VIII.69)}$$

Taking into account the phase shift of $\pi/2$ between ε_r and ε_a (similar to the phase shift between j_r and j_a) and using the symbols $\varepsilon_r = \varepsilon'$, $\varepsilon_a = \varepsilon''$, we can write the permittivity, due to a slowly established polarization, as a complex quantity:

$$\varepsilon^* = \varepsilon' - i\varepsilon''. \qquad \text{(VIII.70)}$$

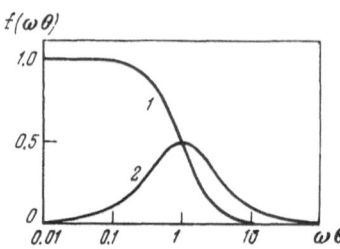

Fig. 224. Dependences $1/[1 + (\omega\theta)^2]$ and $\omega\theta/[1 + (\omega\theta)^2]$ on $\omega\theta$ (curves 1 and 2, respectively).

The real part of Eq. (VIII.70), i.e., the value of ε', represents the capacitative component of the permittivity. During polarization of a dielectric the displacement current, associated with the charging of a capacitance characterized by ε' (the reactive capacitance), leads the voltage by a phase angle $\pi/2$. The imaginary part of the expression (VIII.70), i.e., the value of ε'', represents the active component of the displacement current; this component is in phase with the voltage. The frequency dependences of ε' and ε'' are given by the functions $1/(1 + \omega^2\theta^2)$ and $\omega\theta/(1 + \omega^2\theta^2)$, respectively. These functions are shown graphically in Fig. 224.

The energy dissipated (dielectric losses) per unit volume of a dielectric and per unit time can be found from the relationship

$$p = \frac{i_{am}E_m}{2},\qquad\qquad\text{(VIII.71)}$$

which can be rewritten, using Eq. (VIII.64), in the form

$$p = \frac{1}{2}\frac{\omega^2\theta^2 g}{1 + \omega^2\theta^2}E_m^2 = \frac{\omega^2\theta^2 g}{1 + \omega^2\theta^2}E_{\text{eff}}^2\qquad\text{(VIII.72)}$$

Moreover, employing the relationship between j_{am} and j_{rm}

$$\tan\delta = \frac{i_{am}}{i_{rm}} = \frac{\varepsilon''}{\varepsilon'} = \omega\theta,\qquad\qquad\text{(VIII.73)}$$

we can write Eq. (VIII.72) in the form

$$p = E_{\text{eff}}^2\frac{\omega\varepsilon'}{4\pi}\tan\delta\qquad\qquad\text{(VIII.74)}$$

The quantity $\varepsilon'\tan\delta = \varepsilon''$ is sometimes called the dielectric loss coefficient and the quantity $1/\tan\delta = \varepsilon'/\varepsilon'' = Q$ is called the loss factor. It follows from Eqs. (VIII.73) and (VIII.74) that the losses are small when $\omega\theta \ll 1$. This means that when $\omega \ll 1/\theta$ the active component of the current is small compared with the reactive component and the dielectric is heated only slightly by an alternating field.

This discussion of dielectric losses must be extended to include other types of polarization (in particular, the displacement polarization) and a finite electrical conductivity, which is a property of every real material. As mentioned earlier, the displacement polarization (the electronic and ionic components) is established very rapidly and, therefore, in fields of frequencies up to $\sim 10^{10}$ cps this type of polarization does not give rise to appreciable dielectric losses. The processes associated with charging of the geometrical capacitance (an empty capacitor), which are also completed very rapidly ($< 10^{-10}$ sec), also fail to produce appreciable losses. Apart from the losses − just discussed − associated with the thermal ionic and thermal orientational polarization, there may be losses due to the formation of a space charge. As mentioned in Chap. VII, these processes are slow and give rise to dielectric losses in alternating fields.

The conduction current in a dielectric (ignored so far in the present chapter) is, in fact, the main source of dielectric losses. It is important to stress also (see Chap. VII) that the conduction current redistributes the field. In such cases, the conduction current not only has an active (ohmic) but also a reactive (capacitive) component.

Thus, in a real dielectric the reactive current consists of the capacitative current j_{r1}, due to such processes as charging of the geometrical capacitance or the displacement polarization, and the current j_{r2}, due to the conductivity of a crystal and the relaxational polarization:

$$j_{r_{tot}} = j_{r_1} + j_{r_2}. \tag{VIII.75}$$

Similarly, the active current consists of the current j_a, associated with the relaxational polarization, and the residual conduction current j_{res}:

$$j_{a_{tot}} = j_a + j_{res} \tag{VIII.76}$$

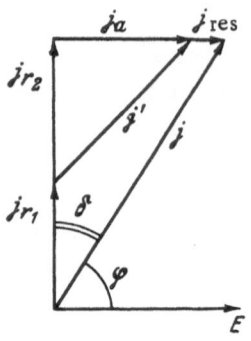

Fig. 225. Vector diagram
of the currents in a real dielectric.

The vector diagram of currents in a real dielectric is shown in
Fig. 225. It is evident from this figure that tan δ is given by the
relationship

$$\tan \delta = \frac{i_a + i_{\text{res}}}{i_{r_1} + i_{r_2}}. \tag{VIII.77}$$

The capacitative component of the permittivity, due to all the
reactive currents, denoted by ε (sometimes known as the total permittivity), is related to the total reactive current by the expression [see Eq. (VIII.65)]*

$$j_{rm_{\text{tot}}} = \frac{\omega_\varepsilon E_m}{4\pi}. \tag{VIII.78}$$

Since $j_{rm_{\text{tot}}} = j_{r_1 m} + j_{r_2 m}$ and

$$j_{r_1 m} = \frac{\omega \varepsilon_\infty}{4\pi} E_m$$

(where ε_∞ is the permittivity which is solely due to the rapidly established polarization), we obtain

$$j_{rm \, \text{tot}} = \omega \left(\frac{\varepsilon_\infty}{4\pi} + \frac{\theta g}{1 + \omega^2 \theta^2} \right) E_m. \tag{VIII.79}$$

---

*The quantity ε is the real or capacitative component of the permittivity, denoted
earlier by ε_r or ε' [see, for example, Eq. (VIII.70)]. For simplicity, we shall now
omit the prime.

Comparing Eqs. (VIII.78) and (VIII.79), we obtain

$$\varepsilon = \varepsilon_\infty + \frac{4\pi\theta g}{1 + \omega^2\theta^2}. \qquad\qquad \text{(VIII.80)}$$

When the voltage is constant ($\omega = 0$) we have

$$\varepsilon_s = \varepsilon_\infty + 4\pi\theta g.$$

At very high frequencies ($\omega \to \infty$) we find that $\varepsilon \approx \varepsilon_\infty$. The dependence $\varepsilon(\omega)$, given by Eq. (VIII.80), is shown graphically in Fig. 226.

We shall now consider dielectric losses in the presence of conduction currents. Using Eq. (VIII.76) for the effective values of the active components of the currents in alternating fields, we obtain:

$$j_{a\,\text{tot.eff}} = \sigma_{\text{res}}E_{\text{eff}} + \frac{\omega^2\theta^2 g}{1 + \omega^2\theta^2}\,E_{\text{eff}} \qquad\qquad \text{(VIII.81)}$$

where σ_{res} is the electrical conductivity estimated from the residual current. Using Eq. (VIII.81), we can describe the specific dielectric losses by the expression

$$p = \left(\sigma_{\text{res}} + \frac{\omega^2\theta^2 g}{1 + \omega^2\theta^2}\right)E^2_{\text{eff}} \qquad\qquad \text{(VIII.82)}$$

Employing Eqs. (VIII.77) and (VIII.80), we can rewrite the last expression in the form

$$p = E^2_{\text{eff}}\frac{\omega\varepsilon}{4\pi}\tan\delta \qquad\qquad \text{(VIII.83)}$$

which is similar to Eq. (VIII.74).

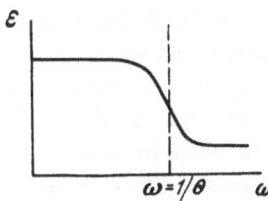

Fig. 226. Frequency dependence of the total permittivity.

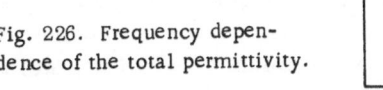

C. Frequency Dependence of Dielectric
Losses. When the voltage is constant, the dielectric losses are
very low and are described by the relationship

$$p = \sigma_{res} E^2. \tag{VIII.84}$$

According to Eq. (VIII.82), the losses increase with increasing ω
approximately proportionally to ω^2 if $\omega\theta \ll 1$. If $\omega\theta \approx 1$, the specific dielectric losses are

$$p = \left(\sigma_{res} + \frac{g}{2}\right) E^2_{eff}. \tag{VIII.85}$$

When the frequency is increased still further the rise of p slows
down: at high frequencies p tends to its maximum value

$$p = (\sigma_{res} + g)\, E^2_{eff}. \tag{VIII.86}$$

The frequency dependence of $\tan \delta$ can be found from Eq.
(VIII.77), which can be reduced to the following form using Eqs.
(VIII.79) and (VIII.81):

$$\tan \delta = \frac{\sigma_{res}(1 + \omega^2\theta^2) + \omega^2\theta^2 g}{\dfrac{\varepsilon_\infty \omega}{4\pi}(1 + \omega^2\theta^2) + \omega g\theta}. \tag{VIII.87}$$

Introducing a special symbol for the difference between the static
and high-frequency values of the permittivity,

$$\varepsilon_s - \varepsilon_\infty = 4\pi\theta g = \Delta_s\varepsilon,$$

and making simple transformations in Eq. (VIII.87), we obtain

$$\tan \delta = \frac{\dfrac{4\pi\sigma_{res}}{\omega\varepsilon_\infty}(1 + \omega^2\theta^2) + \omega\theta\dfrac{\Delta_s\varepsilon}{\varepsilon_\infty}}{1 + \dfrac{\Delta_s\varepsilon}{\varepsilon_\infty} + \omega^2\theta^2}. \tag{VIII.88}$$

Analysis of the above expression is a fairly complex matter,
but we can use it to deduce logical conclusions in two extreme
cases. The first extreme case is obtained when, throughout the
whole range of the frequencies considered, $4\pi\sigma_{res}/\omega$ is large compared with unity and $\Delta_s\varepsilon/\varepsilon_\infty \ll 1$. This case is obtained when the
conductivity is very high (or the frequency ω is low) and the relaxational processes are weak ($\Delta_s\varepsilon \ll 1$). In this case, Eq. (VIII.88)

becomes

$$\tan \delta = \frac{4\pi\sigma_{res}}{\omega\varepsilon_\infty}. \qquad\qquad (VIII.89)$$

Substituting this value of $\tan \delta$ into Eq. (VIII.83), we obtain

$$p = \sigma_{res}E^2_{eff},$$

which shows that, in this limiting case, the dielectric losses in alternating fields are approximately equal to the dc energy losses. These losses are independent of the frequency and are often called the conduction losses.

The other extreme case corresponds to the condition $4\pi\sigma_{res}/\omega \ll 1$ (low conductivity or very high frequencies) but, in this case, the quantity $\Delta_s\varepsilon/\varepsilon_\infty$ is not very small compared with unity (the relaxational processes are strong and the relaxation time fairly large). In this case, Eq. (VIII.88) can be written in the form (after dropping the first term from the numerator)

$$\tan \delta = \frac{\omega\theta\Delta_s\varepsilon}{\varepsilon_s + \varepsilon_\infty\omega_2\theta^2}. \qquad\qquad (VIII.90)$$

If the conductivity is so low that even at low frequencies we have $4\pi\sigma_{res}/\omega \ll 1$, we find that at these frequencies, when $\varepsilon_\infty \omega^2\theta^2 < \varepsilon$, Eq. (VIII.90) becomes much simpler

$$\tan \delta = \omega\theta\frac{\Delta_s\varepsilon}{\varepsilon_s}. \qquad\qquad (VIII.91)$$

According to Eqs. (VIII.90) and (VIII.91), $\tan \delta$ increases first with increasing frequency and, at low frequencies and very low conductivities, its value is approximately proportional to the frequency. When the frequency is increased to values corresponding to the condition $\omega\theta \gg 1$, $\tan \delta$ begins to decrease with increasing frequency. Thus, at some frequency ω_m there must be a maximum of $\tan \delta$. Differentiation of Eq. (VIII.90) with respect to ω gives the value of ω_m and $\tan \delta_{max}$:

$$\omega_m = \frac{1}{\theta}\sqrt{\frac{\varepsilon_s}{\varepsilon_\infty}},$$

$$\tan \delta_{max} = \frac{\Delta_s\varepsilon}{2\sqrt{\varepsilon_s\varepsilon_\infty}}; \qquad\qquad (VIII.92)$$

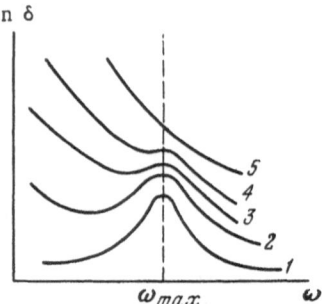

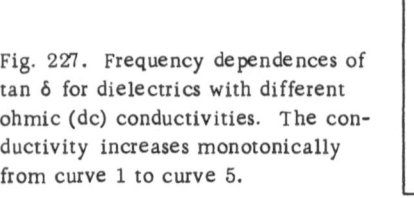

Fig. 227. Frequency dependences of tan δ for dielectrics with different ohmic (dc) conductivities. The conductivity increases monotonically from curve 1 to curve 5.

since, in most cases, ε_s differs little from ε_∞, the frequency corresponding to a maximum of tan δ is close to $1/\theta$.

In real dielectrics we can encounter both these extreme cases as well as all the intermediate ones. If the conductivity of a dielectric is low (the second extreme case), the frequency dependence of tan δ has a sharp maximum. It is evident from Fig. 227 that when the conductivity increases the maximum of tan δ becomes gradually flatter and disappears altogether when the conductivity reaches a sufficiently high value (curve 5); curve 1 represents a dielectric with a low conductivity. Figure 228 shows the dependences of ε, tan δ, and of the specific losses p on the frequency for the same low-conductivity dielectric. It is evident from this figure that the maximum of tan δ, the rise of the losses, and the fall of ε all occur in the same frequency range. The values of ε and p for a dielectric with a high conductivity are independent of frequency, and tan δ of such a dielectric decreases with increasing frequency (the first extreme case).

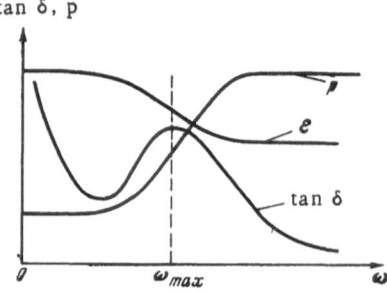

Fig. 228. Frequency dependences of ε, tan δ, and p.

Simple physical interpretations can be given to the relationships just discussed. Thus, the lack of dependence of the specific losses on the frequency (tan $\delta \propto 1/\omega$) in the first extreme case follows from the fact that, in this case, the current in a dielectric subjected to a constant voltage is practically independent of time. The active current is not affected by the application of an alternating voltage. The conduction current varies in phase with the voltage.

In the case of dielectrics with a low conductivity and pronounced relaxational processes, the behavior of tan δ in the frequency range $1/\theta$ is governed by the relationship between the active and reactive components of the current. In the frequency range where tan δ rises, the active component of the current increases more rapidly than the reactive component. At high frequencies tan δ decreases with increasing frequency because the active component of the current is practically independent of the frequency and the reactive component increases proportionally to the frequency and the value of ε remains constant.

Many dielectrics exhibit a weak frequency dependence of tan δ due to the fact that they do not have a single relaxation time but a set of values of relaxation times.

D. Temperature Dependence of Dielectric Losses. The temperature dependence of the loss-angle tangent, written out explicitly, is a fairly cumbersome formula and its analysis is, in general, difficult. However, the temperature dependences in some extreme cases (like the frequency dependences of the losses) yield some physically meaningful conclusions.

We shall give some results of an analysis of Eq. (VIII.87), transformed in such a way so that all the temperature-dependent quantities in this equation are expressed directly in terms of temperature. First of all, we must point out that, in the case of a low value of the dc conductivity, i.e., in the case of the purely relaxational dielectric losses, the temperature dependence of tan δ has a maximum. The temperature at which this maximum of tan δ is observed depends on the frequency. When the frequency is increased, the temperature of the maximum tan δ should shift in the direction of higher temperatures.

This shift can be predicted on the basis of the following simple considerations. When the dc conductivity is low, tan δ depends

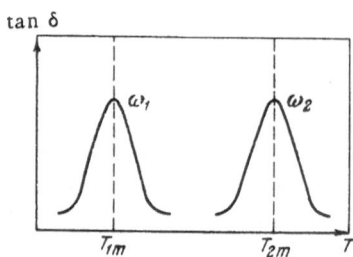

Fig. 229. Temperature dependence of tan δ for relaxational losses (in the absence of dc conductivity).

on the temperature in the same way as the time constant θ. This time constant decreases with increasing temperature. Using the condition for a maximum of tan δ [the first of the two expressions in Eq. (VIII.92)], which applies to this case and can be written in the form

$$(\omega\theta)_m = \sqrt{\frac{\varepsilon_s}{\varepsilon_\infty}}, \tag{VIII.93}$$

we find that, at a fixed frequency, we can alter the temperature so as to obtain such a value of the time constant θ which satisfies the condition (VIII.93). Obviously,

$$\theta_m = \frac{1}{\omega} \sqrt{\frac{\varepsilon_s}{\varepsilon_\infty}}.$$

The temperature dependence of tan δ for the relaxational losses (in the absence of dc conductivity) is shown in Fig. 229. The temperatures T_{1m} and T_{2m} represent the maxima in tan δ at two different frequencies ω_1 and ω_2. Since $\theta_{1m} > \theta_{2m}$, it follows that $\omega_2 > \omega_1$ and the temperature maximum shifts in the direction of higher temperatures when the frequency is increased.

Fig. 230. General temperature dependence of tan δ.

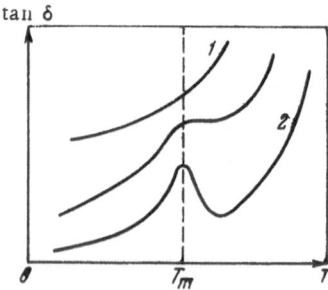

If the relaxational processes in a dielectric are weak, tan δ is given by Eq. (VIII.89), which can be transformed to the following expression by substituting σ_{res} = A exp (−B/T) (see §1 in the present chapter):

$$\tan \delta = \frac{kA}{\omega \varepsilon_\infty} e^{-\frac{B}{T}}, \qquad (VIII.94)$$

where B is the temperature coefficient of the ohmic dc conductivity. In this case, both tan δ and σ_{res} increase exponentially with time.

In general, when a dielectric exhibits both dc conductivity and relaxational losses the temperature dependence of the dielectric losses is of the type shown in Fig. 230. Curve 1 in that figure represents a high-conductivity dielectric, curve 2 a low-conductivity dielectric, and the middle curve represents the intermediate case.

As mentioned earlier, the dielectric losses are governed primarily by the dielectric loss coefficient $\varepsilon \cdot \tan \delta$. In view of this, it is interesting to consider the frequency and temperature dependences of ε. For dielectrics which exhibit the displacement and relaxational polarizations we obtain − using Eq. (VIII.80) and $g\theta \propto 1/T$*:

$$\varepsilon = \varepsilon_\infty + \frac{4\pi A'}{T\,(1 + \omega^2 B'^2)\, e^{\frac{2B''}{T}}},$$

where A', B', and B'' are constants which are independent of temperature.

The last expression shows that at low frequencies or at high temperatures, when $\omega\theta \ll 1$, the second term decreases almost inversely proportionally to the absolute temperature. Conversely, at high frequencies or at moderate temperatures, when $\omega\theta$ is comparable with or greater than unity, the second term increases with increasing temperature. In this case, the permittivity depends on temperature, as shown in Fig. 231a. Curve 1 in this figure represents the behavior of the permittivity at a lower frequency and curve 2 at higher frequency. When the frequency is increased the

---

*In the case of the thermal orientational polarization [see Eq. (VIII.59)], $g = n_0 \mu^2/kT\tau$, where $\tau \sim \theta$; in the case of the thermal ionic polarization, we have $g = n_0 e^2 \delta^2/kT\tau$.

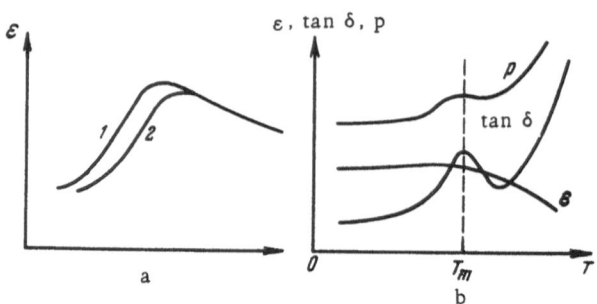

Fig. 231. Temperature dependences of the permittivity at two frequencies (a) and general temperature dependences of the permittivity ε, loss-angle tangent tan δ, and specific losses p (b).

maximum of ε shifts in the direction of higher temperatures. The maximum of ε at a given voltage lies in the range of temperatures in which the product $\omega\theta$ is comparable with unity. Thus, the temperature dependence of ε for a dielectric with pronounced relaxational processes is represented, at a constant voltage, by a falling curve ($\omega = 0$, $\varepsilon = \varepsilon_\infty + 4\pi A/T$), but when the voltage is alternating it is represented by a curve with a maximum whose left-hand side is steeper than its right.

The temperature dependence of the dielectric loss coefficient $\varepsilon \cdot \tan \delta$ (and, therefore, of the magnitude of the dielectric losses) is governed by the product of the dependences $\varepsilon = f(T)$ and $\tan \delta = f(T)$. If the relaxational processes in a dielectric are weak (θ is relatively very small) the losses in which a dielectric are mainly due to conduction. In this case, the losses and tan δ increase with temperature in the same way as the dc conductivity; the permittivity ε is practically independent of the temperature and frequency. In general, when a dielectric exhibits both dc conductivity and a slowly established polarization, the losses and ε depend on the temperature in a more complex manner (Fig. 231b).

The losses in a homogeneous dielectric are directly proportional to the square of the voltage, and the specific losses are proportional to the square of the field intensity. However, we must bear in mind that such a dependence applies only in fields which are not strong enough to produce deviations from Ohm's law or to disturb the proportionality of the polarization to the field. In stronger fields pre-breakdown and breakdown phenomena are observed and these increase dielectric losses, which are then difficult to treat theoretically.

E. Dielectric Losses of Some Linear Dielectrics. Dielectric losses in many crystalline dielectrics are primarily due to their finite electrical conductivity. Only some dielectrics exhibit losses of the relaxational type (for example, the losses due to reorientation of polar groups in ice). The dielectric losses in alkali-halide crystals (NaCl, LiF, KBr, etc.) are governed, in a wide range of temperatures and frequencies, by single vacancies (the concentration of these vacancies is of the order of 10^{15} cm^{-3}), which are responsible for the conductivity of these crystals. However, relatively small relaxational losses, due to reorientation of associated vacancies (at low temperatures there are more of these vacancies: about 10^{17} cm^{-3}), are superimposed on the conduction losses of alkali halide crystals. The temperature dependence of tan δ of LiF crystals is given in Fig. 232 for two frequencies.

In those crystals in which the passage of a current is accompanied by the accumulation of a space charge and the decay of the current with time (this is frequently observed at low temperatures), the dielectric losses are not simply due to the dc conductivity. In such crystals the losses in alternating fields are higher than the losses under a constant voltage.

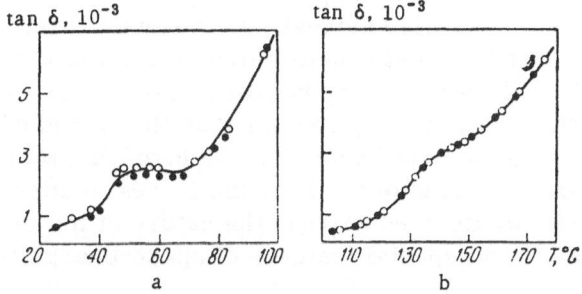

Fig. 232. Temperature dependences of tan δ of LiF single crystals at 1 and 160 kc (a and b, respectively) [13]. Black circles denote values obtained during heating and open circles are values obtained during cooling.

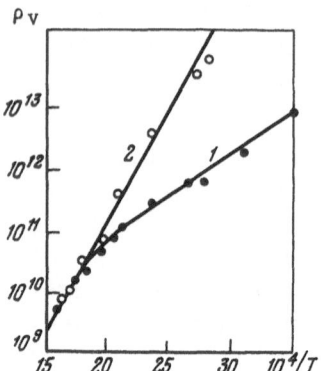

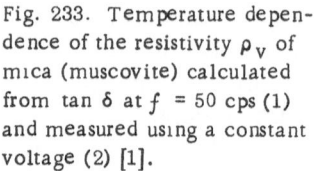

Fig. 233. Temperature dependence of the resistivity ρ_v of mica (muscovite) calculated from $\tan \delta$ at f = 50 cps (1) and measured using a constant voltage (2) [1].

The dc and ac resistivities of mica crystals are equal at high temperatures. This means that at such temperatures there is no appreciable decay of the current with time and the losses are due to the finite conductivity. At low temperatures the dc resistivity of mica is higher than the ac resistivity (Fig. 233).

Quartz crystals have a strongly anisotropic electrical conductivity (mentioned earlier in the present chapter). Consequently, the dielectric losses in quartz are also anisotropic. The ac and dc resistivities along the optic axis are similar, which shows that such losses are governed by the conductivity. At right-angles to the optic axis the ac resistivity is almost an order of magnitude lower than the dc resistivity. On the other hand, measurements of the decay of the current with time show that the time constant characterizing the process of space-charge accumulation is smaller when the field is applied normally to the optic axis. The higher losses in quartz in an alternating field normal to the optic axis are thus of the relaxational type because, under these conditions, the time constant is obviously shorter than the period of the field. In this case, the dielectric losses may be high. In spite of the fact that along both directions of the field the losses in quartz are governed primarily by its conductivity, the nature of the losses is different because of the special features of space-charge accumulation.

§ 5. Dielectric Losses in Ferroelectrics

We have mentioned earlier that dielectric losses are governed by the nature of the polarization processes in a crystal. The establishment of polarization in linear dielectrics ($P = \alpha E$, $\alpha = $ const) is relatively simple. The dc polarization of such dielectrics depends exponentially on the duration of application of the field. The time for establishment of the polarization θ (θ occurs in the argument of the exponential function) is quite definite for each type of polarization and is independent of the value of the field. A definite equilibrium state (in the thermodynamic sense) of polarization P exists in a linear dielectric at any value of the field E.

Experimental investigations of the polarization in ferroelectrics are described in detail in Chap. VII. Generalizing the experimental data, we can say that, in contrast to linear dielectrics, ferroelectrics do not obey the simple linear dependence of the polarization on the field and a ferroelectric crystal is in a state of equilibrium only when $P = 0$ (if $E = 0$) or if $|P| \approx P_s$ (if $E \neq 0$), where P_s is the spontaneous polarization.* The transition from the state with $P = P_s$ to the state with $P = -P_s$ (polarization reversal) involves two processes: nucleation and growth of domains of a given sign. The time for establishment of the total polarization depends on the field E.

In sinusoidal electric fields the special features of the polarization of ferroelectrics give rise to a displacement current with a complex nonsinusoidal dependence on time (because of the strongly nonlinear dependence of the polarization on the field) and a very high amplitude (the active component of this current is responsible for the dielectric losses). The amplitude, period, and harmonic composition of the displacement current all depend strongly on the sinusoidal field E. Therefore, a simple mathematical description of the polarization process cannot include the establishment of polarization and the generation of dielectric losses in ferroelectrics.

A. Dielectric Losses in Ferroelectrics Subjected to Weak Fields. In weak fields ferroelectrics are polarized (in the first approximation) similarly to linear dielectrics (Chap. VII). Although domain processes do take place, they do not give rise to a hysteretic dependence of the polarization on the field, and the motion of domain walls can be regarded as

---

* The polarization P can be represented as the sum of the spontaneous polarization P_s and of the additional elastic polarization ΔP ($P = P_s + \Delta P$). Usually, $\Delta P \ll P_s$ and, therefore, we may assume that $|P| \approx P_s$.

elastic. It follows that the principal relationships governing the weak-field losses in ferroelectrics are similar to the relationships deduced by Debye for the relaxational polarization of linear dielectrics.

The establishment of polarization and dielectric losses in ferroelectrics subjected to weak fields has been investigated in a wide range of frequencies from 10 to 10^{13} cps (§5, Chap. VII). As with the polarization of linear dielectrics, the maximum of the loss-angle tangent of ferroelectrics lies in a relatively narrow range of frequencies in which the permittivity exhibits dispersion. This is the range of frequencies in which the time for establishment of the polarization is comparable with the period of the external electric field ($\omega\theta \sim 1$). Outside this dispersion region, i.e., at frequencies which satisfy the inequalities $1 \ll \theta\omega \ll 1$, the loss-angle tangent is very small. The behavior of the loss-angle tangent can thus be used to deduce information on the nature of the polarization of a crystal.

The frequency dependence of the loss-angle tangent of a barium titanate single crystal is shown in Fig. 234. The value of tan δ, which is a few hundredths at radio frequencies, increases to 0.12-0.13 at 1 Gc, remains constant up to 56 Gc, and then increases very rapidly. The rise of tan δ at 1 Gc and the corresponding increase of the dispersion of ε (Fig. 184a) are usually explained by the inertia of domain walls. All the mechanisms associated with domain motion cease to be active in the polarization processes above a frequency corresponding to the frequency of natural oscillations of domain walls.

This point of view is supported also by the observation that the permittivity and losses decrease with increasing constant elec-

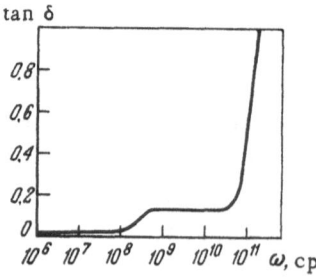

Fig. 234. Frequency dependence of tan δ of a BaTiO$_3$ single crystal at room temperature.

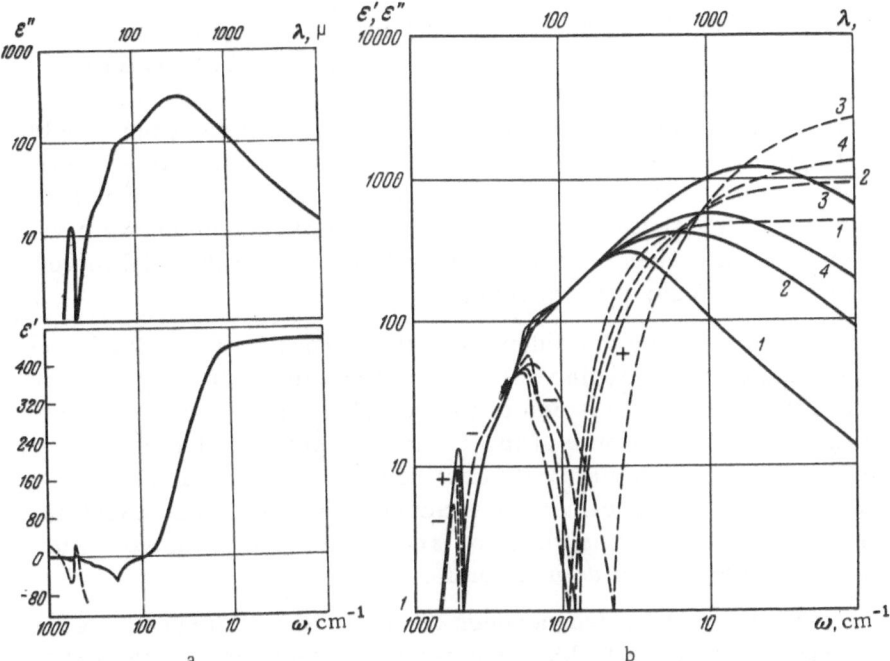

Fig. 235. Dependences of the real (ε') and imaginary (ε'') parts of the permittivity of polycrystalline BaTiO$_3$ on the wavelength (wave number): a) at room temperature (dashed curve shows the dependence of $10\varepsilon'$); b) at various temperatures (°C): 1) 45; 2) 80; 3) 110; 4) 140 (dashed curves represent ε' and continuous curves represent ε'') [15].

tric field. This drop in ε and tan δ is also due to the "exclusion" of domains by the polarizing field.

The rise of tan δ at still higher frequencies (10^{11}–10^{12} cps) is associated with special features of the polarization which are due to the lattice vibrations (Chap. VI). Investigations of the polarization and losses in this range of frequencies are very interesting because they provide information on the nature of the spontaneous polarization. The behavior of the polarization and of the losses in the millimeter and infrared range of wavelengths is shown in Fig. 235a for the tetragonal modification of BaTiO$_3$.

The value of tan δ for the low-temperature modifications of BaTiO$_3$ (orthorhombic and trigonal) is considerable and amounts to 0.2–0.5. The value of tan δ for the paraelectric (cubic) modification of $BaTiO_3$ is small up to 10^{10} cps. This can be regarded as an additional proof of the domain nature of the 10^9 cps losses in the tetragonal modification. The paraelectric modification of barium titanate exhibits a dispersion of ε and a rise of $\tan \delta$ only at frequencies above 24 Gc. The loss-angle tangent of this modification is 0.1. The dispersion of ε in this range of frequencies is associated with the lattice vibrations. Figure 235b shows how ε' and ε'' vary when the temperature is altered in the millimeter and infrared range of wavelengths.

The frequency dependences of the permittivity and the loss-angle tangent of triglycine sulfate are given in §5 of Chap. VII, in connection with a discussion of the polarization of ferroelectrics (Fig. 184b). A maximum of $\tan \delta$ and a dispersion of ε are observed at a frequency (10^6-10^7 cps) considerably lower than the relaxation frequency of barium titanate. The dielectric losses in triglycine sulfate in this range of frequencies are associated with the forward growth of domain nuclei.

The frequency dependences of the dielectric losses have been investigated in weak fields for other ferroelectrics. In particular, guanidine aluminum sulfate crystals exhibit a loss maximum ($\tan \delta \sim 0.1$) in the frequency range 10^6-10^7 cps; this maximum is obviously of domain origin.

The value of $\tan \delta$ of ferroelectrics increases and passes through a maximum when the temperature approaches the ferroelectric phase transition point. The maximum of $\tan \delta$ of $BaTiO_3$ lies a little below the Curie temperature (120°C). This is because even in weak fields the losses near the Curie temperature increase due to a high domain mobility. The losses decrease when the temperature is lowered because this reduces the domain mobility.

An analytic expression for the frequency and temperature dependences of $\tan \delta$ of a ferroelectric in a weak field can be obtained by considering the dynamics of the domain-wall motion. Analyses of the domain polarization in weak fields have been carried out by Kittel [23], Fatuzzo [22], and others. For example, the model suggested by Fatuzzo relates the relaxation of the polarization to the oscillations of the front wall of a ferroelectric domain (i.e., to the growth of a domain or its forward motion). The expression obtained by Fatuzzo for the complex permittivity is

$$\Delta\varepsilon = \frac{C}{(\omega_0^2 - \omega^2) + j\omega\gamma}, \qquad \text{(VIII.95)}$$

where $C = 32\pi P_s^2 (t_d/t_s) dm$; P_s is the spontaneous polarization; ω_0 and γ are the dynamic parameters of an oscillating wall; t_d and t_s are, respectively, the period of motion of a domain wall and polarization reversal time; d is the thickness of a crystal; m is the effective mass per unit area of the domain wall. In order that the polarization should be of the relaxational type, we must assume that $\omega_0 \ll \gamma$. If the angular frequency of the field is $\omega \ll \omega_0$, we find that

$$\Delta\varepsilon = \frac{C / \omega_0^2}{1 + j\omega (\gamma / \omega_0^2)}. \qquad \text{(VIII.96)}$$

This expression is similar to the Debye equation describing the relaxational polarization of linear dielectrics if we assume that $C/\omega_0^2 \equiv \Delta\varepsilon'$ and $\gamma/\omega_0^2 \equiv \tau$.

B. Dielectric Losses in Ferroelectrics Subjected to Strong Fields. The nonlinear dependence of the polarization on the field, observed when strong fields are applied to ferroelectrics, gives rise to dielectric hysteresis loops. In this case, the state of polarization of a crystal depends on the amplitude and frequency of the field. In general, in a wide range of frequencies (right up to 10^8 cps) we can select a field which produces complete alignment of the domains along the field (Fig. 188).

The dielectric losses in ferroelectrics subjected to strong fields can be due to various causes but the main losses are due to dielectric hysteresis. The magnitude of these losses is proportional to the area of the hysteresis loop.

In linear dielectrics subjected to a sinusoidal external electric field the displacement current varies sinusoidally because of the linear dependence of the polarization on the field. A phase shift between the displacement current and its reactive component, measured in terms of angles (§4 in the present chapter), determines the dielectric losses in linear dielectrics. The tangent of this angle, known as the loss-angle tangent, is equal to the ratio of the active (i.e., dissipated) and reactive components of the power.

The nonlinear dependence of the polarization of ferroelectrics on the field gives rise to higher harmonics of the displacement current (its fundamental frequency is equal to the frequency
of the applied field). The coefficient of the nonlinear distortions of
the displacement current $k = (i_2^2 + i_3^2 + .../i_1)^{1/2}$, where i_1, i_2, i_3, ...
are the amplitudes of the first, second, and third, etc. harmonics
of the displacement current, usually amounts to a few tenths. We
can easily see that, in this case, the concept of the loss-angle tangent loses its meaning. However, in order to retain the analogy
with the loss characteristics of ordinary dielectrics, it is usual
to speak of the loss-angle tangent of ferroelectrics even in strong
fields. The value of tan δ is defined, as in the case of ordinary
dielectrics, as the ratio of the energy lost in one period to the
maximum electric energy stored in a crystal. This quantity is
usually of the order of unity.

In some cases it is important to know the behavior of a ferroelectric crystal in a sinusoidal field at the fundamental frequen-

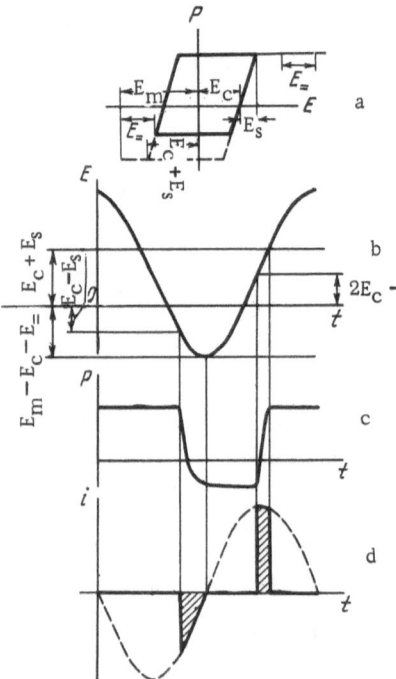

Fig. 236. Time dependences of the polarization P(c) and of the displacement current
i(d) for an idealized hysteresis loop (a) in
an electric field E (b) [10].

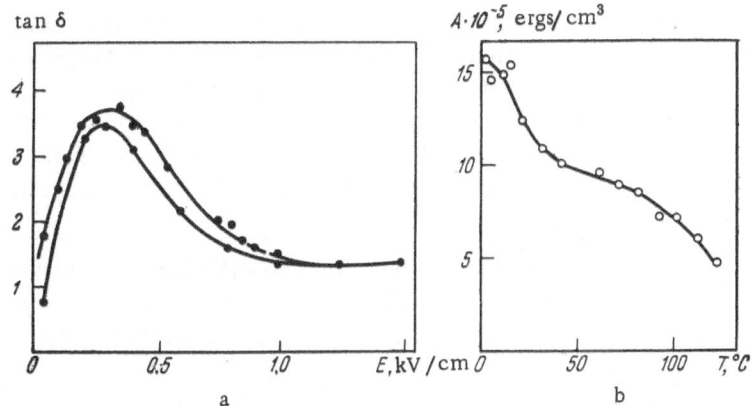

Fig. 237. Dielectric losses in some ferroelectrics: a) $\tan \delta = f(E_\sim)$ for tri-glycine sulfate, determined using the first harmonic [10]; b) temperature dependence of the hysteresis losses in $BaTiO_3$ [2].

cy, i.e., the amplitude and phase of the first (fundamental) harmonic of the displacement current. In this case, the higher harmonics are ignored and the concept of the loss–angle tangent has still the same meaning as in ordinary dielectrics.

Considering an idealized hysteresis loop (Fig. 236a) and the corresponding displacement current (Fig. 236d) we can, using Fourier series expansions, find the amplitude and phase of the first (fundamental) harmonic of the current and the loss–angle tangent. The relationships obtained depend on the parameters of the idealized hysteresis loop: the amplitude of the external field E_m, the coercive field E_c, the saturation field E_s, the spontaneous polarization P_s, and the constant bias field $E_=$ (if such a field is applied). In the most interesting case, when $E_s = 0$ and $E_m = E_c$ (i.e., when the hysteresis loop is rectangular), the amplitude of the first (fundamental) harmonic of the displacement current is $i_1 \propto 2f P_s$, the phase shift between this current and the external field is zero, and $\tan \delta = \infty$. This means that a ferroelectric with a rectangular hysteresis loop behaves as an ohmic resistance.

Hysteresis loops of real dielectrics (for example, barium titanate and triglycine sulfate) are nearly rectangular. The losses in such crystals are very high. Figure 237a shows the dependence of $\tan \delta$ of a triglycine sulfate single crystal on the amplitude of an external field E (this dependence was deduced from the first harmonic of the displacement current). A maximum of the value of tan δ is found in a field E = 0.27 kV/cm and it is equal to 3.8. The values of tan δ which lie to the left of this maximum correspond to "unsaturated" hysteresis loops, i.e., to such amplitudes of E in which the domain polarization has not yet been fully established.

The absolute value of the losses due to dielectric hysteresis can be expressed in terms of the hysteresis loop area. The work done by an external field in altering the polarization of a ferroelectric by an amount dP (the voltage is U) is given by

$$dA = sU\,dP,\qquad\qquad\text{(VIII.97)}$$

where s is the area of the electrodes of the ferroelectric capacitor. During cyclic polarization reversal the total work per cycle is proportional to the hysteresis loop area and equal to

$$A = s \oint U\,dP.\qquad\qquad\text{(VIII.98)}$$

In the case of rectangular hysteresis loops we can easily obtain the following approximate expression for the calculation of the power loss for hysteresis at an external field frequency f:

$$P = Af = f \oint E \cdot d \cdot s \cdot dP \cong 4P_s E_c f d s,\qquad\text{(VIII.99)}$$

where P_s is the spontaneous polarization; E_c is the coercive field; d is the thickness of a crystal. The value of the specific power loss is

$$W = \frac{P}{v} = 4P_s E_c f,\qquad\qquad\text{(VIII.100)}$$

where $v = d \cdot s$ is the volume being subjected to polarization reversal.

The power lost due to dielectric hysteresis is transformed into heat. The value of this power may be fairly high. Thus, for example, in a barium titanate single crystal, d = 0.44 mm thick with an electrode area s = 1 mm², the power dissipated at 1 Mc is over 160 W. The losses due to dielectric hysteresis do not have

a maximum in their frequency dependence but they increase monotonically with frequency (provided complete polarization reversal takes place at all the investigated frequencies).

We can easily see from Eq. (VIII.100) that the losses due to
hysteresis depend on the temperature in the same way as the spontaneous polarization P_s and the coercive field E_c. The losses decrease when the Curie temperature T_c is approached from the ferroelectric range of temperatures. This is illustrated in Fig. 237b,
which gives the temperature dependence of the losses in $BaTiO_3$.

The electric field energy dissipated in a dielectric is converted into heat. The much lower losses of ferroelectrics in the
paraelectric range of temperatures can be used in self-stabilization of their temperature. In this self-stabilization process, a
ferroelectric is heated by the high losses in the ferroelectric state
until it reaches the temperature of its phase transition to the paraelectric modification. In the paraelectric range of temperatures
the losses are lower and the crystal begins to cool. However,
this cooling is temporary and the crystal again warms up because
of the higher losses in the ferroelectric state. Such self-stabilization has been observed in various ferroelectrics (triglycine sulfate, $BaTiO_3$, and others).

The ability of ferroelectrics to maintain their temperature
in external fields can be used in special ferroelectric thermostats.

References

1.   N. P. Bogoroditskii and V. N. Malyshev, Zh. Tekh. Fiz., 4:1036 (1934).
2.   V. A. Bokov, Dissertation [in Russian], Institut Kristallografii AN SSSR, Moscow
     (1959).
3.   A. T. Venderovich and B. I. Lapkin, Zh. Éksp. Teor. Fiz., 9:46 (1939).
4   A. A. Vorob'ev, Physical Properties of Ionic Crystalline Dielectrics [in Russian],
     Vol. 1, Tomsk (1960).
5.   A. A. Vorob'ev and E. K. Zavadovskaya, Electric Strength of Solid Dielectrics
     [in Russian], GTTI, Moscow (1956).
6.   B. V. Gorelik and V. T. Dmitriev, Zh. Tekh. Fiz., 18:333 (1948).
7.   A. F. Ioffe, K. D. Sinel'nikov, and B. M. Gokhberg, Zh. Russk Fiz-Khim. Obshch.,
     Chast' Fiz., 58:105 (1926).
8.   V. M. Gurevich and I. S. Zheludev, Kristallografiya, 5:805 (1960).
9.   I. S. Zheludev, Zh. Tekh. Fiz., 24:1467 (1954).
10. I. S. Zheludev, V. V. Gladkii, L. Z. Rusakov, and I. S. Rez, Izv. Akad. Nauk
     SSSR, Ser. Fiz., 22:1465 (1958).
11. F. Seitz, Modern Theory of Solids, McGraw-Hill, New York (1940).

12.   A. F. Ioffe, Izv. Petrogradsk. Politekhn. Inst. (1915).

13.   B. N. Matsonashvili and G. I. Skanavi, in: Physics of Dielectrics (Proc. Second All-Union Conf., Moscow, 1958) [in Russian], Izd. AN SSSR, Moscow (1960).

14.   N. F. Mott and R. W. Gurney, Electronic Processes in Ionic Crystals, Clarendon Press, Oxford (1948).

15.   V. N. Murzin and A. I. Demeshina, Fiz. Tverd. Tela, 6:182 (1964).

16.   G. I. Skanavi, Physics of Dielectrics [in Russian], GTTI, Moscow (1949).

17.   G. I. Skanavi, Dielectric Polarization and Losses in High-Permittivity Glasses and Ceramics [in Russian], Gosénergoizdat, Moscow (1952).

18.   G. I. Skanavi, Physics of Dielectrics in Strong Fields [in Russian], Fizmatgiz, Moscow (1958).

19.   W. Franz, "Theorie des rein elektrischen Durschlags festen Isolatoren", Ergeb. Exakt. Naturwiss., 27:1 (1953).

20a. Ya. I. Frenkel', Zh. Éksp. Teor. Fiz., 8:1292 (1938).

20b. Ya. I. Frenkel', Élektrichestvo, No. 8, 5 (1947).

21.   P. Debye, Polar Molecules, Chemical Catalog Co., New York (1929).

22.   E. Fatuzzo, J. Appl. Phys., 32:1571 (1962).

23.   C. Kittel, Phys. Rev., 83:458 (1951).

24.   H. H. Poole, Phil. Mag., 42:488 (1921).

25   W. F. Brown, Jr., "Dielectrics," in: Handbuch der Physik (ed. by S. Flügge), Vol. 17, Part 1, Springer Verlag, Berlin (1956), p. 1.

26.   W. Känzig, "Ferroelectrics and antiferroelectrics," Solid State Phys., 4:1 (1957).

27.   C. Kittel, Introduction to Solid State Physics, 3rd ed , Wiley, New York (1966).

28.   H. Fröhlich, Theory of Dielectrics, Dielectric Constant, and Dielectric Loss, 2nd ed., Clarendon Press, Oxford (1958).

29.   A. von Hippel (ed.), Dielectric Materials and Applications, Wiley, New York and the Technology Press, Massachusetts Inst. of Technology (1954).

Chapter IX

Piezoelectric and Electrostrictive Properties

Introduction

The piezoelectric and electrostrictive properties of crystals (particularly the former) have important practical applications. Piezoelectric crystals are so widely used in hydroacoustics, electroacoustics, communications, and measurement techniques that it is impossible to imagine the operation of many electrical and electronic devices and measuring instruments without piezoelectric crystals. The new methods for growing piezoelectric crystals and the development of a large number of piezoelectric devices have given rise to whole branches of industry. Many monographs and papers have been written on the piezoelectric properties and on devices based on them.

We shall not attempt to give here a complete review of the work done on piezoelectricity. We shall restrict ourselves to the most important aspects of this complex physical phenomenon. Our attention will be concentrated on the interrelationship between the mechanical, electrical, and (to a lesser extent) thermal properties of crystals. In discussing the piezoelectric properties of ferroelectric crystals, we shall pay special attention to the role played by the spontaneous polarization in piezoelectricity.

In spite of the extensive work done on piezoelectricity, the theoretical analysis of the nature of the piezoelectric effect has not progressed beyond the thermodynamic description of the effect. In fact, it would be difficult to find more than two or three serious investigations of the nature of the piezoelectric effect, molecular models, etc. This situation is due to the general lag of the theory of spontaneous polarization behind practical applications of dielectrics.

Electrostrictive phenomena in crystals have been investigated less thoroughly than piezoelectricity and they are much less frequently employed in practical applications. Only recently, in connection with the work on ferroelectrics (which exhibit strong electrostriction), has the interest in these phenomena increased. Investigations of electrostriction (particularly the electrostriction of linear dielectrics) remain difficult because of the lack of apparatus for measuring infinitesimally small displacements.

§1. Crystallography of the Piezoelectric Effect

A. Direct Piezoelectric Effect. When mechanical stresses or strains are applied to a crystal (even when they can be described by symmetrical polar tensors) it is found that special polar directions may appear in a crystal. Hence, we have the possibility (at least in principle) of the appearance of electric polarization (along these special polar directions) due to applied stresses or strains.

Moreover, mechanical stresses or strains in spontaneously polarized crystals can alter the magnitude or direction of P_s which which can be regarded as a polarization due to mechanical forces. Such symmetry considerations were used by Pierre and Jacques Curie in their search for electric polarization in crystals caused by mechanical stresses or strains. The possibility of such polarization was demonstrated experimentally by the Curie brothers in 1880 and the phenomenon was later called the piezoelectric effect. It was demonstrated later still that some crystals could be deformed by the application of an electric field. The first of these phenomena is now known as the direct piezoelectric effect and the second is called the converse (sometimes also reciprocal or inverse) effect.

It has been established experimentally that the piezoelectric effect is linear. The components of the polarization vector $\mathbf{P}$ are related linearly to the components of the tensors of mechanical stresses $[t_{ij}]$ or strains $[r_{ij}]$:

$$P_1 = d_{111}t_{11} + d_{112}t_{12} + d_{113}t_{13} + d_{121}t_{21} + d_{122}t_{22} + d_{123}t_{23} + d_{131}t_{31} + {} $$
$$+ d_{132}t_{32} + d_{133}t_{33}, \qquad (IX.1)$$

where d_{ijk} are constant coefficients which are called piezoelectric strain coefficients or piezoelectric moduli. The expressions for the components P_2 and P_3 are similar. Using the summation rules for repeated indices, Eq. (IX.1) can be written in the form

$$P_i = d_{ijk} t_{jk} . \qquad (IX.2)$$

The coefficients d, of which there are 27, form a third-rank tensor whose components transform in accordance with the law

$$d'_{ijk} = C_{il} C_{jm} C_{kn} d_{lmn},$$
$$d_{ijk} = C_{li} C_{mj} C_{nk} d'_{lmn}, \qquad (IX.3)$$

where C are the components of the matrix of direction cosines used in the transformation of coordinates.

If we assume that the tensor $[t_{ij}]$ is symmetrical, the coefficients d_{ijk} should obey the condition

$$d_{ijk} = d_{ikj}. \qquad (IX.4)$$

It is convenient to express the piezoelectric strain coefficients d using two indices by employing a single index for the tensor $[t_{ij}]$, in accordance with the rules

$$t_{11} = t_1; \quad t_{22} = t_2; \quad t_{33} = t_3;$$
$$t_{32}, t_{23} = t_4; \quad t_{31}, t_{13} = t_5; \quad t_{12}, t_{21} = t_6. \qquad (IX.5)$$

When this is done the tensor $[d_{ijk}]$ of Eq. (IX.2)

	t_{11}	t_{22}	t_{33}	t_{32}	t_{23}	t_{31}	t_{13}	t_{12}	t_{12}
P_1	d_{111}	d_{122}	d_{133}	d_{132}	d_{123}	d_{131}	d_{113}	d_{112}	d_{121}
P_2	d_{211}	d_{222}	d_{233}	d_{232}	d_{223}	d_{231}	d_{213}	d_{212}	d_{221}
P_3	d_{311}	d_{322}	d_{333}	d_{332}	d_{323}	d_{331}	d_{313}	d_{312}	d_{321}

$(IX.6)$

transforms into

	t_1	t_2	t_3	t_4	t_5	t_6
P_1	d_{11}	d_{12}	d_{13}	d_{14}	d_{15}	d_{16}
P_2	d_{21}	d_{22}	d_{23}	d_{24}	d_{25}	d_{26}
P_3	d_{31}	d_{32}	d_{33}	d_{34}	d_{35}	d_{36} .

$(IX.7)$

Comparison of the coefficients d_{ijk} of the tensor in Eq. (IX.6) and of the coefficients d_{ij} of the matrix in Eq. (IX.7)* yields the following relationships

$$d_{111} = d_{11}; \quad d_{122} = d_{12}; \quad d_{133} = d_{13};$$

$$d_{132}, \ d_{123} = \frac{1}{2} d_{14}; \quad d_{131}, \ d_{113} = \frac{1}{2} d_{15}; \quad d_{112}, \ d_{121} = \frac{1}{2} d_{16};$$

$$d_{211} = d_{21}; \quad d_{222} = d_{22}; \quad d_{233} = d_{23}; \quad d_{232}, \ d_{223} = \frac{1}{2} d_{24};$$

$$d_{231}, \ d_{213} = \frac{1}{2} d_{25}; \quad d_{212}, \ d_{221} = \frac{1}{2} d_{26}; \quad d_{311} = d_{31};$$

$$d_{322} = d_{32}; \quad d_{333} = d_{33}; \quad d_{323}, \ d_{332} = \frac{1}{2} d_{34}; \quad d_{313}, \ d_{331} = \frac{1}{2} d_{35}; \qquad \text{(IX.8)}$$

$$d_{312}, \ d_{321} = \frac{1}{2} d_{36}.$$

In compact form Eq. (IX.7) can be written thus:

$$P_i = d_{ij} t_j \qquad (i = 1, 2, 3; \qquad j = 1, 2, \ldots, 6). \qquad \text{(IX.9)}$$

The symmetry of a crystal imposes restrictions on the form of the matrix of the piezoelectric coefficients d_{ij}. These restrictions are usually allowed for analytically.

For this purpose, as in all other similar cases in crystal physics, we write the expressions for the coefficients d_{ij} after transformation of the coordinate system in accordance with some symmetry element of a crystal. After such a transformation we may find that:

a) the coefficient is equal to itself ($d'_{ij} = d_{ij}$);

b) the coefficient is equal to itself numerically but its sign is opposite to that before the transformation ($d'_{ij} = -d_{ij}$);

c) the transformed coefficient is equal to the algebraic combination of the original coefficients.

The restrictions on the form of the (d_{ij}) matrix, corresponding to these three cases, are as follows:

a) $d'_{ij} = d_{ij}$ means that the presence of this nonzero coefficient does not contradict the symmetry of a crystal and the coefficient remains at its original place in the table;

---

* The coefficients d_{ij} in Eq. (IX.7) will be called the coefficients of the matrix; if they are to be considered as coefficients of a third-rank tensor we must use three indices.

b) $d'_{ij} = -d_{ij}$ means that this equality is compatible with the symmetry of a crystal only for one value of d_{ij}, namely, $d_{ij} \equiv 0$;

c) if the coefficient is a combination of other coefficients we can simply use the new coefficient. In this case, the value of the coefficient is compatible with the symmetry of the crystal.

The results of an analytic treatment of the form of the matrices of the piezoelectric strain coefficients of noncentrosymmetrical classes of crystals are given in Table 46 [9] (all coefficients vanish for the centrosymmetrical classes). It is evident from this table that in the case of 20 classes the matrix of the coefficients d_{ij} of Eq. (IX.7) has at least one nonzero coefficient.

In the nineteen-forties, Shubnikov showed that piezoelectric properties should be exhibited not only by crystals but also by textures consisting of "particles" of different nature and ordered in accordance with symmetry laws. In the simplest case, such textures (known as piezoelectric textures) consist of oriented crystallites forming a polycrystalline aggregate. In general, textures can have any (including noncrystallographic) symmetry. However, if we restrict our discussion to the effects described by second-rank tensors, the symmetry groups of textures are identical with the symmetry groups of continuous uniform media [1]. The only textures of intrinsic interest are those corresponding to the noncrystallographic groups, i.e., the limiting Curie groups (Chap. I): $\infty/m \cdot \infty{:}m$; $\infty/\infty{:}2$; $m \cdot \infty{:}m$; $\infty{:}m$; $\infty \cdot m$; $\infty{:}2$; ∞.

The limiting groups $\infty/m \cdot \infty{:}m$, $m \cdot \infty{:}m$, and $\infty{:}m$ are of no interest from the point of view of the possibility of piezoelectric polarization because the textures of these groups are centrosymmetrical and, on the basis of the same considerations which apply to crystals, piezoelectric polarization cannot appear in centrosymmetrical textures.

An analytic treatment shows that the piezoelectric coefficients are also all equal to zero for the textures belonging to the symmetry group $\infty/\infty{:}2$, although this texture is not centrosymmetrical. In this case, there is no orientation of the symmetry elements of a crystal and of the symmetrical tensors $[t_{ij}]$ or $[r_{ij}]$ for which this texture would become a polar group. For any orientation of applied forces the symmetry group of such a texture becomes $\infty{:}2$, which does not have even polar directions (although it has no center of symmetry or inversion center).

TABLE 46. Form of the (d_{ij}) Matrices for Crystals [9]

Notation

· zero coefficient

● nonzero coefficient

●—● equal coefficients

●—○ coefficients numerically equal but opposite in sign

◎ coefficients opposite in sign and numerically equal
    to twice the heavy-dot coefficient to which it is joined

The number of different nonzero coefficients is given
in parentheses after each matrix

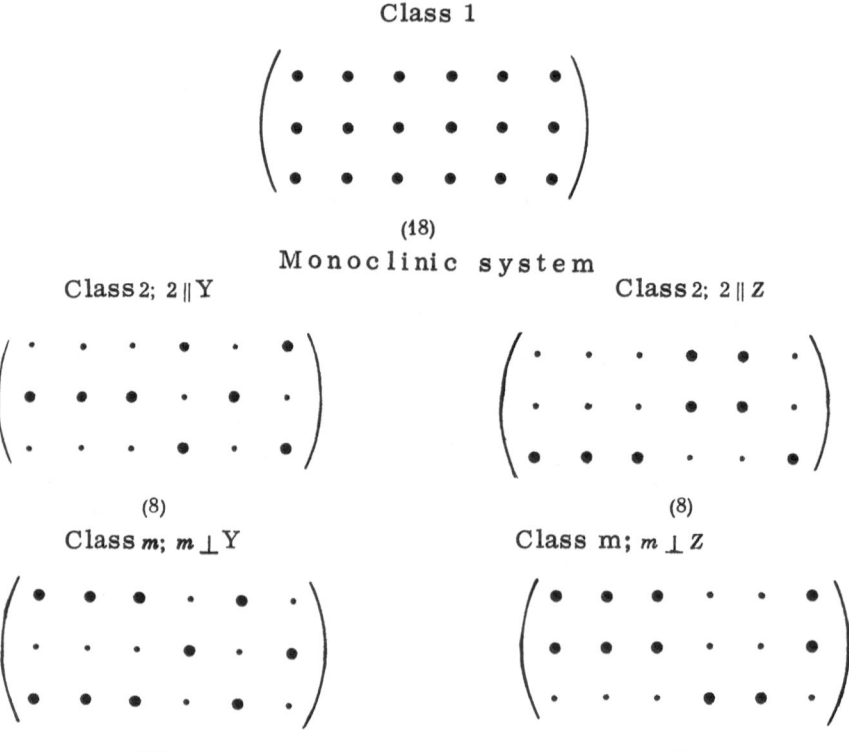

Triclinic system
Class 1

(18)

Monoclinic system

Class 2; 2 ∥ Y   Class 2; 2 ∥ Z

(8)   (8)

Class *m*; *m* ⊥ Y   Class m; *m* ⊥ Z

(10)   (10)

TABLE 46 (Continued)

Trigonal system

Class 3

(6)

Class 3 : 2

(2)

Class 3·m; m ⊥ X

(4)

Class 3·m; m ⊥ Y

(4)

Hexagonal system

Class 6

(4)

Class 6·m

(3)

Class 6 : 2

(1)

Class 3 : m

(2)

Class m·3 : m; m ⊥ X

(1)

Class m·3 : m; m ⊥ Y

(1)

TABLE 46 (Continued)

Orthorhombic system

Class 2 : 2   Class 2·*m*

(3)   (5)

Tetragonal system

Class 4   Class 4̄

(4)   (4)

Class 4 : 2   Class 4·*m*

(1)   (3)

Class 4̄·*m*; 2 ‖ *X*, *Y*

(2)

Cubic system
Classes 3/4 and 3/2

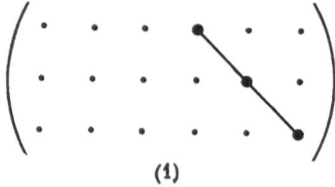

(1)

Textures belonging to the symmetry groups $\infty\cdot m$, $\infty{:}2$, and ∞ have nonzero coefficients in the piezoelectric tensors. The matrices of these coefficients are listed in Table 47.

We have mentioned earlier that the electric polarization P in the direct piezoelectric effect may be due to a strain described by the tensor $[r_{ij}]$. In this case, the equation for the piezoelectric effect can be written in the form

$$P_i = e_{ijk} r_{jk}, \tag{IX.10}$$

where e_{ijk} are new piezoelectric coefficients (stress coefficients) in a third-rank tensor. Mechanical stresses $[t_{ij}]$ and strains $[r_{ij}]$ can be related not only to the value of the polarization P but also to the value of the resultant electric field E. This gives two new equations for the direct piezoelectric effect

$$E_i = -g_{ijk} t_{jk}, \tag{IX.11}$$
$$E_i = -h_{ijk} r_{jk}, \tag{IX.12}$$

where g_{ijk} and h_{ijk} are piezoelectric coefficients in third-rank tensors.

TABLE 47. Form of the (d_{ij}) Matrices for Some Piezoelectric Textures

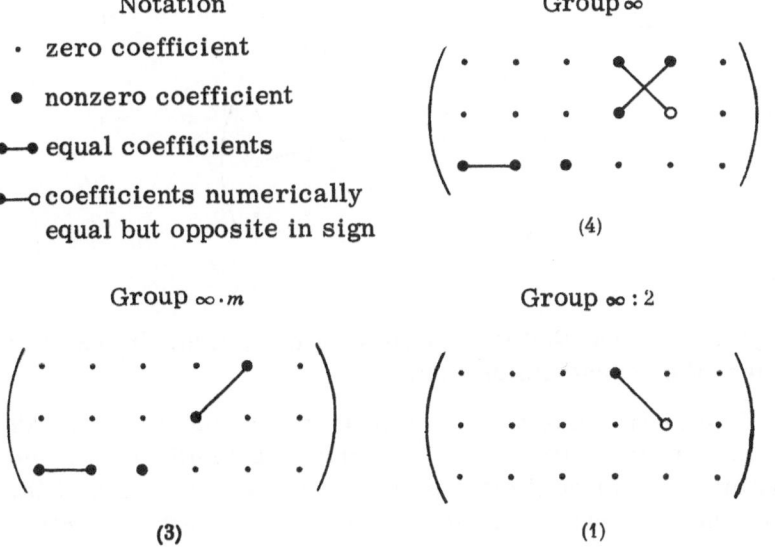

Notation

· zero coefficient

• nonzero coefficient

•—• equal coefficients

•—o coefficients numerically equal but opposite in sign

Group ∞

(4)

Group $\infty\cdot m$

(3)

Group $\infty:2$

(1)

TABLE 48. Form of the (e_{ij}) and (h_{ij}) Matrices for Crystal
Classes with (e_{ij}) and (h_{ij}) Different from (d_{ij})

Notation

·   zero coefficient

●   nonzero coefficient

$\left.\begin{array}{c} \circ\!\!-\!\!\circ \\ \bullet\!\!-\!\!\bullet \end{array}\right\}$ equal coefficients

○——●   coefficients numerically equal but opposite in sign

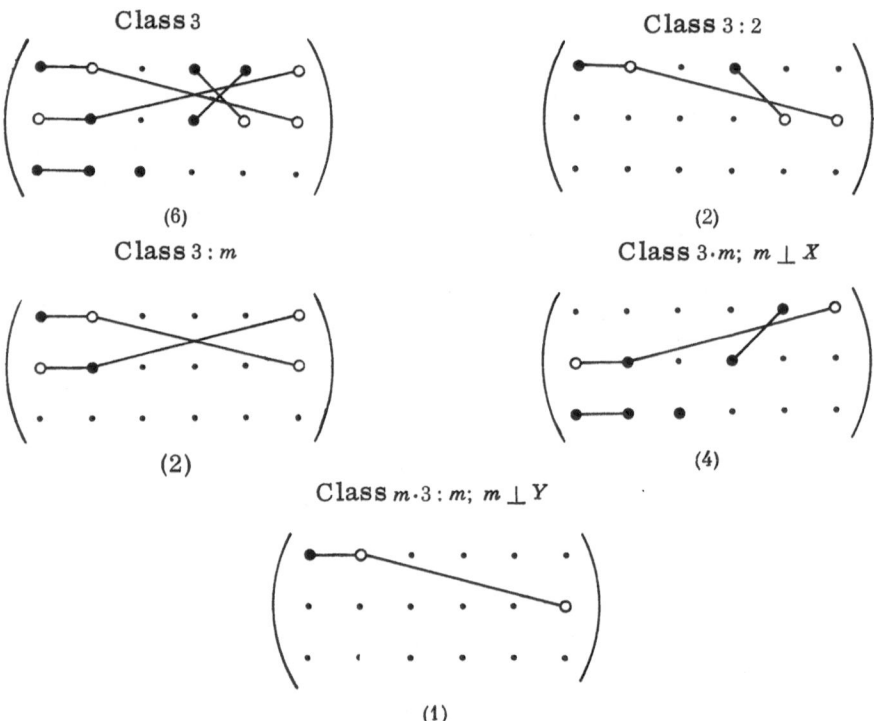

Class 3

(6)

Class 3 : 2

(2)

Class 3 : m

(2)

Class 3·m; $m \perp X$

(4)

Class m·3 : m; $m \perp Y$

(1)

We can show that the coefficients d, e, g, and h are related
through other constants of a crystal.

The matrices of the coefficients g_{ij} are identical with the
matrices of the coefficients d_{ij} given in Tables 46 and 47. However, for some crystal classes the form of the matrices of the
coefficients e_{ij} and h_{ij} will differ from the matrices of the coefficients

d_{ij}; this happens when the strain is described by a matrix whose coefficients are related to the components of the tensor $[r_{ij}]$ by $r_1 = r_{11}$, $r_2 = r_{22}$, $r_3 = r_{33}$, $r_4 = 2r_{23}$, $r_5 = 2r_{31}$, $r_6 = 2r_{12}$. The matrices for such crystals are given in Table 48. When we go over from two indices to three indices in the coefficients e and h, we must drop the factors $1/2$ from formulas similar to Eq. (IX.8). The matrices of the coefficients d_{ij}, g_{ij}, e_{ij}, and h_{ij} for piezoelectric textures are identical.

B. Converse Piezoelectric Effect. We shall now consider the converse effect. In the converse effect the strain in a crystal r_{ij} is proportional to the applied field E:

$$r_{ij} = d_{kij}E_k, \tag{IX.13}$$

where d_{kij} are piezoelectric strain coefficients. In the next section we shall show that the converse effect is the thermodynamic consequence of the direct piezoelectric effect. Taking into account other possible relationships between the strain $[r_{ij}]$ and stress $[t_{ij}]$ tensors, on the one hand, and the electric field E and the polarization P, on the other, we obtain three additional equations for the converse piezoelectric effect*

$$r_{ij} = g_{kij}P_k, \qquad t_{ij} = e_{hij}E_h, \qquad t_{ij} = h_{hij}P_k, \tag{IX.14}$$

where e_{kij}, g_{kij}, and h_{kij} are the piezoelectric coefficients introduced earlier and described by third-rank tensors.

It follows from $r_{ij} = r_{ji}$ that the piezoelectric strain coefficients d_{kij} of Eq. (IX.13) obey the relationship $d_{kij} = d_{kji}$. Moreover, using single-index notation for the strain tensor, we can obtain the piezoelectric strain coefficients d_{kij} with two indices. In this case, Eq. (IX.13) assumes the form

$$r_i = d_{ji}E_j.$$

or, in the matrix form,

---

*The symbol P in the equations for the converse piezoelectric effect is understood to represent the value of free charge σ_0, related to the electric induction (flux density) by the expression $D = 4\pi P = 4\pi \sigma_0$; $P = D/4\pi$.

	E_1	E_2	E_3
r_1	d_{11}	d_{21}	d_{31}
r_2	d_{12}	d_{22}	d_{32}
r_3	d_{13}	d_{23}	d_{33}
r_4	d_{14}	d_{24}	d_{34}
r_5	d_{15}	d_{25}	d_{35}
r_6	d_{16}	d_{26}	$d_{36}.$

$$\text{(IX.15)}$$

Thus, in the converse piezoelectric effect the strain and the field are related by the same coefficients which are used in the equations relating the polarization and stress. In other words, Eqs. (IX.13) and (IX.15) describe the converse piezoelectric effect if the direct effect is described by

$$P_j = d_{ji}t_i. \qquad \text{(IX.16)}$$

Similar comments can be made about the coefficients e, g, and h.

The crystal symmetry imposes definite restrictions on the matrices of the coefficients d_{ji}, e_{ji}, g_{ji}, and h_{ji}, which have been discussed in detail in the preceding section and are given in Tables 46 and 48.

The converse piezoelectric effect is exhibited only by those classes of crystals in which the direct effect can be observed. The converse effect, like the direct effect, cannot exist in centrosymmetrical crystal classes. It follows that Tables 46-48 can be used for the direct and converse piezoelectric effects.

The converse effect can be exhibited not only by crystals but also by piezoelectric textures.

§2. Thermodynamics of Piezoelectric Properties of Crystals

The piezoelectric effect is a phenomenon in which mechanical strains or stresses in a dielectric give rise to an electric polarization which is directly proportional to such strains or stresses; this proportionality also applies to the converse effect. Thus, the piezoelectric effect exhibits a linear relationship between electrical and mechanical quantities, i.e., it is a linear effect.

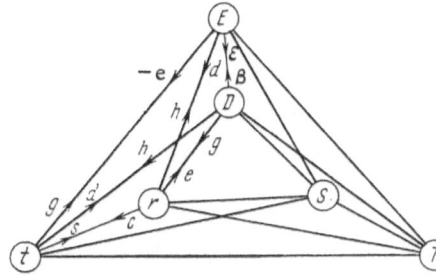

Fig. 238. Schematic representation of the relationships between thermal, electrical, and mechanical properties of a crystal.

The direct and converse piezoelectric effects are fairly complex physical phenomena. The complexity is mainly due to a multifaceted relationship of the piezoelectric effect with elastic, thermal, and electrical properties of crystals. The formal relationships between the quantities representing thermal, electrical, and mechanical properties of crystals are shown in Fig. 238. The relationships involving the piezoelectric effect (on the left-hand side of this figure) are indicated by thick lines.

In general, thermal, elastic, and electrical properties of crystals can be described by thermodynamic functions: internal energy, free energy, Gibbs function, and so on (Table 49), which are functions of independent variables given at the corners of the triangles in Fig. 238 (t, E, T, r, D, S). Each of these functions has three independent variables, one of which represents the mechanical state of a crystal, another — the electrical state, and the third — the thermal state.

A. Equations for the Piezoelectric Effect. We shall describe the piezoelectric effect using the internal energy of a crystal, U, which is a function of r_i, D_m, S (i = 1, 2, ..., 6; m = 1, 2, 3). Thus,

$$dU = T\,dS + t_i\,dr_i + \frac{1}{4\pi} E_m\,dD_m. \qquad (IX.17)$$

Hence,

$$t_i = \frac{\partial U}{\partial r}; \; E_m = 4\pi \frac{\partial U}{\partial D_m}; \; dT = \frac{\partial U}{\partial S}. \qquad (IX.18)$$

Expanding as series the functions describing the mechanical stresses t and an electric field E, and assuming adiabatic conditions (S = const), we obtain

TABLE 49. Thermodynamic Functions Used in Descriptions of Electrical, Mechanical, and Thermal Properties of Crystals

Function	Independent variables	Differential relationships
Internal energy U	$r_i,\ D_m,\ S$	$dU = t_i dr_i + E_m \dfrac{dD_m}{4\pi} + TdS$
Free energy $A = U - ST$	$r_i,\ D_m,\ T$	$dA = t_i dr_i + E_m \dfrac{dD_m}{4\pi} - SdT$
Enthalpy $H = U - r_i t_i - E_m \dfrac{D_m}{4\pi}$	$t_i,\ E_m,\ S$	$dH = -r_i dt_i - \dfrac{D_m}{4\pi} dE_m + TdS$
Elastic enthalpy $H_1 = U - r_i t_i$	$t_i,\ D_m,\ S$	$dH_1 = -r_i dt_i + E_m \dfrac{dD_m}{4\pi} + TdS$
Electric enthalpy $H_2 = U - E_m \dfrac{D_m}{4\pi}$	$r_i,\ E_m,\ S$	$dH_2 = t_i dr_i - \dfrac{D_m}{4\pi} dE_m + TdS$
Gibbs function $G = U - r_i t_i - \dfrac{E_m D_m}{4\pi} - ST$	$t_i,\ E_m,\ T$	$dG = -r_i dt_i - \dfrac{D_m}{4\pi} dE_m - SdT$
Elastic Gibbs function $G_1 = U - r_i t_i - ST$	$t_i,\ D_m,\ T$	$dG_1 = -r_i dt_i + E_m \dfrac{dD_m}{4\pi} - SdT$
Electric Gibbs function $G_2 = U - E \dfrac{D_m}{4\pi} - ST$	$r_i,\ E_m,\ T$	$dG_2 = t_i dr_i - \dfrac{D_m}{4\pi} dE_m - SdT$

$$t_1 = \left(\frac{\partial t_1}{\partial r_1}\right)_{D,\,S} dr_1 + \left(\frac{\partial t_1}{\partial r_2}\right)_{D,\,S} dr_2 + \cdots + \left(\frac{\partial t_1}{\partial r_6}\right)_{D,\,S} dr_6 +$$

$$+ 4\pi \left(\frac{\partial t_1}{\partial D_1}\right)_{r,\,S} dD_1 + 4\pi \left(\frac{\partial t_1}{\partial D_2}\right)_{r,\,S} dD_2 + 4\pi \left(\frac{\partial t_1}{\partial D_3}\right)_{r,\,S} dD_3;$$

$$\dotfill \quad \text{(IX.19)}$$

$$t_6 = \left(\frac{\partial t_6}{\partial r_1}\right)_{D,\,S} dr_1 + \left(\frac{\partial t_6}{\partial r_2}\right)_{D,\,S} dr_2 + \cdots + \left(\frac{\partial t_6}{\partial r_6}\right)_{D,\,S} dr_6 +$$

$$+ 4\pi \left(\frac{\partial t_6}{\partial D_1}\right)_{r,\,S} dD_1 + 4\pi \left(\frac{\partial t_6}{\partial D_2}\right)_{r,\,S} dD_2 + 4\pi \left(\frac{\partial t_6}{\partial D_3}\right)_{r,\,S} dD_3;$$

$$E_1 = \left(\frac{\partial E_1}{\partial r_1}\right)_{D,\,S} dr_1 + \left(\frac{\partial E_1}{\partial r_2}\right)_{D,\,S} dr_2 + \cdots + \left(\frac{\partial E_1}{\partial r_6}\right)_{D,\,S} dr_6 +$$

$$+ 4\pi \left(\frac{\partial E_1}{\partial D_1}\right)_{r,\,S} dD_1 + 4\pi \left(\frac{\partial E_1}{\partial D_2}\right)_{r,\,S} dD_2 + 4\pi \left(\frac{\partial E_1}{\partial D_3}\right)_{r,\,S} dD_3;$$

$$\dotfill \quad \text{(IX.20)}$$

$$E_3 = \left(\frac{\partial E_3}{\partial r_1}\right)_{D,\,S} dr_1 + \left(\frac{\partial E_3}{\partial r_2}\right)_{D,\,S} dr_2 + \cdots + \left(\frac{\partial E_3}{\partial r_6}\right)_{D,\,S} dr_6 +$$

$$+ 4\pi \left(\frac{\partial E_3}{\partial D_1}\right)_{r,\,S} dD_1 + 4\pi \left(\frac{\partial E_3}{\partial D_2}\right)_{r,\,S} dD_2 + 4\pi \left(\frac{\partial E_3}{\partial D_3}\right)_{r,\,S} dD_3.$$

The partial derivatives of the stresses t_i with respect to the strains r_i, which occur in Eq. (IX.19), are the elastic moduli (also known as rigidity moduli) c_{ij} in generalized Hooke's law

$$r_{ij} = s_{ijkl}t_{kl}; \quad t_{ij} = c_{ijkl}r_{kl};$$

the matrices of the elastic constants (also known as compliance constants) s and of the elastic moduli c are given in Table 50. According to the differentiation conditions, the elastic moduli c_{ij} should be measured at constant electric induction D and constant entropy S, which will be indicated by writing c_{ij} with superscripts D and S. The partial derivatives of the $(\partial t / \partial D)_{r,\,S}$ type represent increases in the mechanical stresses caused by the electric induction; these increases relieve the strains in the crystal (r = 0). We shall denote these partial derivatives by the symbols h_{ij}: they are third-rank tensors, which describe the converse piezoelectric effect. The quantities h_{ij} always represent the ratio of the mechanical stresses to the electric induction and, therefore, we shall not use the superscript r with h_{ij}. Assuming that a crystal expands under the action of D, we find that in order to maintain r = const we need compressive ("negative") mechanical stresses. In view

TABLE 50. Form of the (s_{ij}) and (c_{ij}) Matrices

Notation

·   zero coefficient

●   nonzero coefficient

●—● equal coefficients

○—● coefficients numerically equal but opposite in sign

⊙   for s: coefficient numerically equal to twice the heavy-dot coefficient to which it is joined; for c: coefficient numerically equal to the heavy-dot coefficient to which it is joined

× $2(s_{11} - s_{12})$ for s and $\frac{1}{2}(c_{11} - c_{12})$ for c

Triclinic System
Both classes

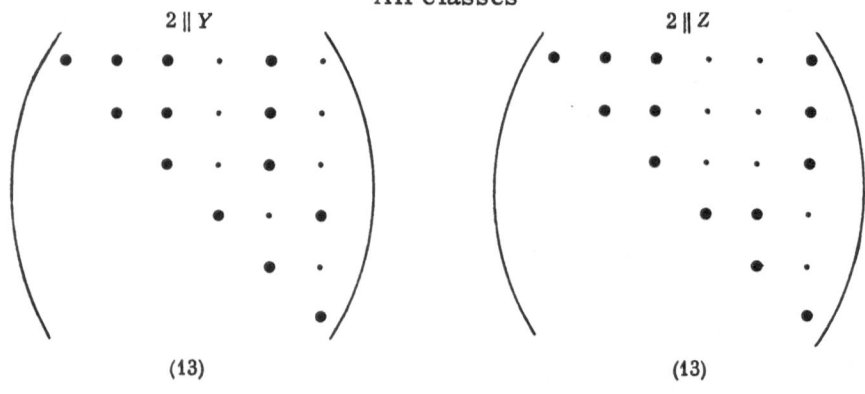

(21)

Monoclinic System
All classes

2 ∥ Y

(13)

2 ∥ Z

(13)

TABLE 50 (Continued)

Orthorhombic System
All classes

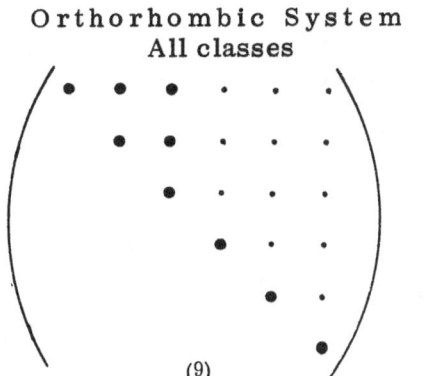

(9)

Cubic System
All classes

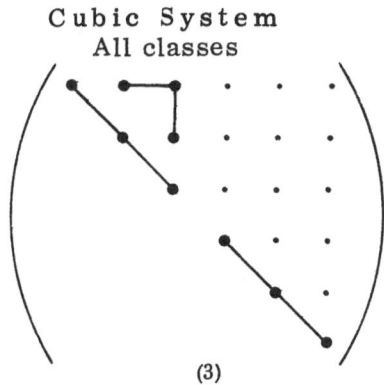

(3)

Tetragonal System

Classes 4; $\bar{4}$; 4 : m

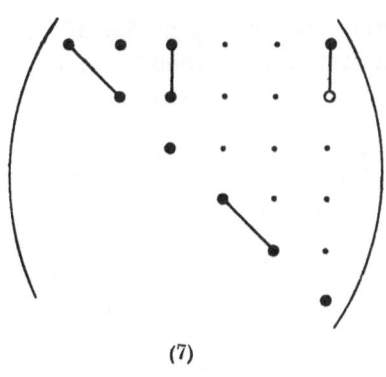

(7)

Classes $m \cdot 4 : m$; $\bar{4} \cdot m$; 4 : 2; $4 \cdot m$

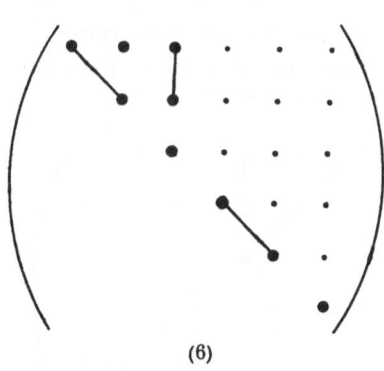

(6)

Trigonal System

Classes 3; $\bar{6}$

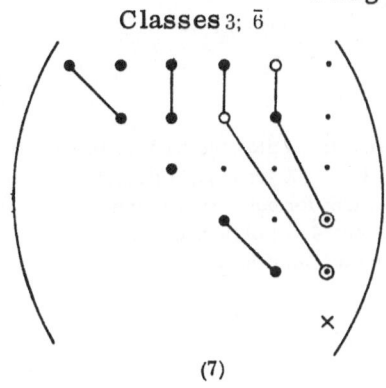

(7)

Classes 3 : 2; $\bar{6} \cdot m$: $3 \cdot m$

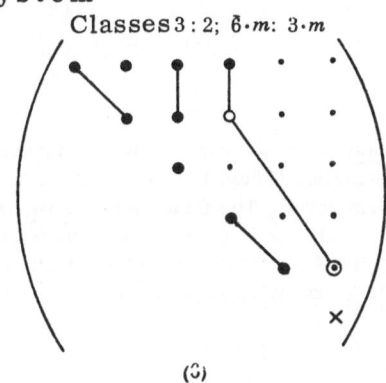

(3)

TABLE 50 (Continued)

Hexagonal System

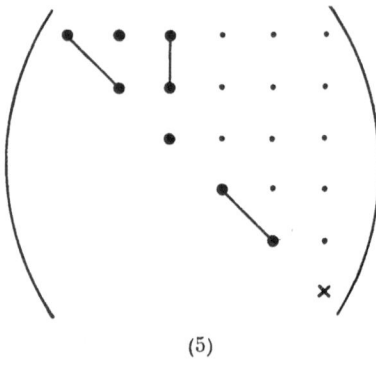

(5)

Isotropic media with sym-
metries described by groups
$\infty/m\cdot\infty{:}m$ and $\infty/\infty{:}2$

Uniform continuous media of
symmetries described by groups
$m\cdot\infty{:}m$, $\infty{:}m$, $\infty\cdot m$, $\infty{:}2$, ∞

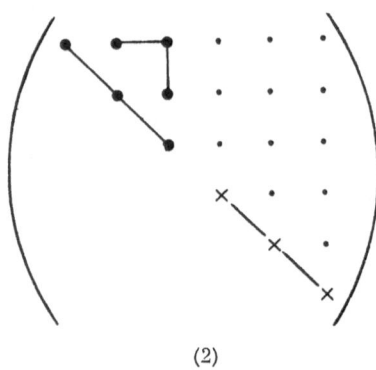

(2)

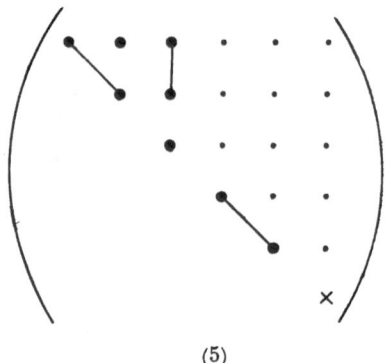

(5)

Note: 1) In the case of monoclinic crystals it is usual to employ the $2\,\|\,Y$ (or $m\perp Y$) orientation (Chap. I, Table 5); however, sometimes the $2\,\|\,Z$ (or $m\perp Z$) orientation is employed. The table gives the (s_{ij}) and (c_{ij}) matrices for both orientations.

2) The form of the (s_{ij}) and (c_{ij}) matrices for the $\bar{4}\cdot m$ class is the same for the X and Y axes coinciding with axes 2 (assumed orientation) and for the X and Y axes lying in planes of symmetry.

of this we shall use the minus sign in front of h_{ij} in equations such as (IX.19).

We shall now consider the system of equations (IX.20). Since Eq. (IX.17) is a total differential, we find that differentiation of t_j with respect to D_n and of E_n with respect to r_j gives

$$- h_{ij} = 4\pi \left(\frac{\partial t_j}{\partial D_n} \right)_{r,\,s} = \left(\frac{\partial E_n}{\partial r_j} \right)_{D,\,s} = 4\pi \left(\frac{\partial^2 U}{\partial r_j \partial D_n} \right) \equiv$$

$$\equiv 4\pi \left(\frac{\partial^2 U}{\partial D_n \partial r_j} \right) \qquad (n = 1,\, 2,\, 3). \tag{IX.21}$$

Thus, terms of the $(\partial E_n / \partial r_j)_{D,S}$ type in Eq. (IX.20) are, like the corresponding quantities in Eq. (IX.19), third-rank tensors but, in this case, they describe the direct piezoelectric effect: they represent the increase in the electric field necessary to maintain constant electric induction in a crystal when this increase is due to mechanical strain. Terms of the type

$$4\pi \left(\frac{\partial E_m}{\partial D_n} \right)_{r,\,s} = \beta^{r,\,s}$$

are components of a tensor which is a reciprocal of the permittivity tensor ε_{nm} . They are obtained from the tensor ε_{nm} using the expressions

$$\beta = \frac{(-1)^{m+n} \Delta^{m,n}}{\Delta},$$

where Δ is a determinant of the components of the tensor ε_{nm}; $\Delta^{m,n}$ are minors obtained by crossing out the m-th row and the n-th column in the determinant.

It follows from our discussion that the six equations in (IX.19) and the three equations in (IX.20) can be written in the form

$$t_1 = c_{11}^{D,\,S} r_1 + c_{12}^{D,\,S} r_2 + \cdots + c_{16}^{D,\,S} r_6 - \frac{1}{4\pi} h_{11}^{S} D_1 - \frac{1}{4\pi} h_{21}^{S} D_2 - \frac{1}{4\pi} h_{31}^{S} D_3;$$

$$\cdots\cdots\cdots\cdots\cdots\cdots\cdots\cdots\cdots\cdots\cdots\cdots\cdots\cdots\cdots \tag{IX.22}$$

$$t_6 = c_{61}^{D,\,S} r_1 + c_{62}^{D,\,S} r_2 + \cdots + c_{66}^{D,\,S} r_6 - \frac{1}{4\pi} h_{11}^{S} D_1 - \frac{1}{4\pi} h_{21}^{S} D_2 - \frac{1}{4\pi} h_{31}^{S} D_3;$$

$$E_1 = \beta_{11}^{r,\ S} D_1 + \beta_{12}^{r,\ S} D_2 + \beta_{13}^{r,\ S} D_3 - h_{12}^{S} r_1 - \cdots - h_{16}^{S} r_6;$$

$$\cdots\cdots\cdots\cdots\cdots\cdots\cdots\cdots\cdots\cdots\cdots\cdots\cdots\cdots\cdots \quad \text{(IX.23)}$$

$$E_3 = \beta_{31}^{r,\ S} D_1 + \beta_{32}^{r,\ S} D_2 + \beta_{33}^{r,\ S} D_3 - h_{9}^{S} r_1 - h_{32}^{S} r_2 - \cdots - h_{36}^{S} r_6.$$

The last two equations can be written in a generalized form:

$$t_j = c_{j1}^{D} r_1 + c_{j2}^{D} r_2 + \cdots + c_{j6}^{D} r_6 - \frac{1}{4\pi} h_{1j} D_1 - \frac{1}{4\pi} h_{2j} D_2 - \frac{1}{4\pi} h_{3j} D_3, \quad \text{(IX.24)}$$

$$E_m = \beta_{m1}^{r} D_1 + \beta_{m2}^{r} D_2 + \beta_{m3}^{r} D_3 - h_{m1} r_1 - h_{m2} r_2 - \cdots - h_{m6} r_6. \quad \text{(IX.25)}$$

In the last two equations, the superscript S, indicating constancy of the entropy, is omitted from the coefficients c, h, and β because these coefficients are usually measured under adiabatic conditions. However, we must remember that, in general, the adiabatic and isothermal values of these coefficients are different.

Thus, our analysis of the internal energy U has yielded relationships between mechanical stresses and electric polarization in the converse piezoelectric effect as well as relationships between electric fields and mechanical strains in the direct effect. In general (Fig. 238), mechanical stresses may be related to electric polarization or induction in the direct effect and electric fields can be related to mechanical strains in the converse effect. Moreover, mechanical stresses may be related not to electric polarization or induction but to electric fields, and so on. It is evident that the two pairs of electrical and mechanical quantities in the direct and converse effects can be related in eight ways. However, it follows from our discussion that the same piezoelectric coefficient relates two different quantities in the direct and converse effects and, therefore, we obtain only four groups of equations. One of these groups comprises Eqs. (IX.24) and (IX.25).

Another (second) group of equations can be obtained using the enthalphy H (Table 49), which depends on the variables t_i, E_m, and S. When the quantities r_i and D_m are expanded into series and it is assumed that S = const (adiabatic conditions), we obtain derivatives of the type $\partial r/\partial E$ and $\partial D/\partial t$. The derivative $(\partial r/\partial E)_{t,S}$ describes a strain in a free crystal generated by the application of an electric field (the converse piezoelectric effect) and the derivative $(\partial D / \partial t)_{E, S}$ represents a change in the electric induction under the influence of mechanical stresses (the direct effect). The last two derivatives are equal because an increase in the enthalpy H is a total differential:

$$\frac{1}{4\pi}\left(\frac{\partial D_m}{\partial t_j}\right)_{E, S} = \left(\frac{\partial r_j}{\partial E_m}\right)_{t, S} = \frac{\partial^2 H}{\partial t_j \partial E_m} = d_{mj}. \qquad \text{(IX.26)}$$

The coefficients d_{mj} are the piezoelectric strain coefficients (also known as piezoelectric moduli), defined in the preceding section.

The generalized form of the second group of equations is written in the form

$$r_i = s_{i1}^E t_1 + s_{i2}^E t_2 + \cdots + s_{i6}^E t_6 + d_{1i}E_1 + d_{2i}E_2 + d_{3i}E_3, \qquad \text{(IX.27)}$$

$$D_m = 4\pi d_{m1}t_1 + 4\pi d_{m2}t_2 + \cdots + 4\pi d_{m6}t_6 + e_{m1}^t E_1 + e_{m2}^t E_2 + e_{m2}^t E_3; \text{(IX.28)}$$

here, s_{ij} are the elastic (complicance) constants; the superscripts E and t indicate conditions under which the relevant coefficients are measured; the index indicating constancy of the entropy (S = const), which indicates that a given coefficient is measured under adiabatic conditions, is omitted from the above equations. Using an elastic enthalpy H_1 and an electrical enthalpy H_2, we can obtain the other two groups (third and fourth) of equations for the piezoelectric effect under adiabatic conditions:

$$r_1 = s_{i1}^D t_1 + s_{i2}^D t_2 + \cdots + s_{i6}^D t_6 + \frac{1}{4\pi} g_{1i}D_1 + \frac{1}{4\pi} g_{2i}D_2 + \frac{1}{4\pi} g_{3i}D_3, \qquad \text{(IX.29)}$$

$$E_m = \beta_{m1}^t D_1 + \beta_{m2}^t D_2 + \beta_{m3}^t D_3 - g_{m1}t_1 - g_{m2}t_2 - \cdots - g_{m6}t_6, \qquad \text{(IX.30)}$$

and

$$t_j = c_{j1}^E r_1 + c_{j2}^E r_2 + \cdots + c_{j6}^E r_6 - e_{1j}E_1 - e_{2j}E_2 - e_{3j}E_3, \qquad \text{(IX.31)}$$

$$D_n = 4\pi e_{n1}r_1 + 4\pi e_{n2}r_2 + \cdots + 4\pi e_{n6}r_6 + e_{n1}^r E_1 + e_{n2}^r E_2 + e_{n3}^r E_3. \qquad \text{(IX.32)}$$

We shall now discuss the physical meaning of two piezoelectric coefficients g and e, which appear in the third and fourth group of equations. The piezoelectric coefficient g represents a strain caused by an electric induction D in the converse effect [Eq. (IX.29)]. However, in the direct effect [Eq. (IX.30)] the same coefficient represents an open–circuit voltage produced by a given mechanical stress. The piezoelectric stress coefficient e (known also as the piezoelectric constant) gives a stress which must be applied to a crystal in the converse effect in order to retain its strain-free state in a given external field E [Eq. (IX.31)]. In the direct piezoelectric effect the coefficient e represents an electric induction D generated by a mechanical stress [Eq. (IX.32)].

All the equations describing the piezoelectric effect [Eqs. (IX.24), (IX.25), (IX.27)-(IX.32)] can be written in a simpler form in those cases when the electrical parameters D and E in the direct piezoelectric effect are governed solely by mechanical strains and stresses and when the mechanical properties in the converse effect are governed solely by electrical parameters. Thus, considering only the intrinsic piezoelectric phenomena, we obtain

$$
\begin{aligned}
t_j &= -\frac{1}{4\pi} h_{nj} D_n, \\
E_m &= -h_{mi} r_i, \\
r_i &= d_{ni} E_n, \\
D_m &= 4\pi d_{mi} t_i, \\
r_i &= \frac{1}{4\pi} g_{ni} D_n, \\
E_m &= g_{mj} t_j, \\
t_j &= -e_{mj} E_m, \\
D_m &= 4\pi e_{mi} r_i.
\end{aligned}
\tag{IX.33}
$$

The system of equations (IX.33) can be simplified still further. The fourth and eighth of these equations describe the direct piezoelectric effect. If we consider only the electric fields generated in the direct effect (it is assumed that there are no external fields and this makes it possible to describe the direct effect by simpler equations), we can interpret D in the fourth and eighth equations as the charge which compensates the piezoelectric polarization and ensures the absence of an electric field E in a capacitor containing a piezoelectric material:

$$
E = 0. \tag{IX.34}
$$

Equation (IX.34) is satisfied when

$$
\frac{D}{4\pi} = P \qquad (D = E + 4\pi P), \tag{IX.35}
$$

where P is the piezoelectric polarization.

The field E can be made to vanish by the adsorption of external charges on a crystal or by the use of electrodes. In the first case, external charges which compensate the piezoelectric polarization are usually charges which exist in the ambient medium (for example, in air) and which are adsorbed by the surface of a crystal. If electrodes are used, the condition E = 0 can be satisfied by short-circuiting the electrodes of a polarized sample. In this case, the charges of opposite signs, which appear on the electrodes because of the electrostatic induction, are neutralized and the remaining free charges ensure that the condition (IX.35) is satisfied.*

Thus, bearing in mind the conclusions reached in our discussion, the system of equations (IX.33) can be rewritten in the form

$$
\begin{aligned}
t_j &= \frac{1}{4\pi}\, h_{nj} D_n, \\
E_m &= -\, h_{mi} r_i, \\
r_i &= d_{ni} E_n, \\
P_m &= d_{mi} t_i, \\
r_i &= \frac{1}{4\pi}\, g_{ni} D_n, \\
E_m &= -\, g_{mj} t_j, \\
t_j &= -\, e_{mj} E_m, \\
P_n &= e_{ni} r_i.
\end{aligned}
\tag{IX.36}
$$

The system of equations (IX.33) can be used, bearing in mind the relationship between the elastic coefficients c and s as well as the relationship between D and E, to obtain the following expressions which relate the various piezoelectric coefficients. These expressions are of the form

---

* As already mentioned in Chap. VII, the normal component of the polarization P is numerically equal to the surface density of the bound charge: $P_n = \sigma_b$. When a piezoelectric material is placed in a parallel-plate capacitor and the direction of polarization is perpendicular to the electrodes we obtain $P = \sigma_b$. In the case of an electrically short-circuited crystal, the free and bound charges are numerically equal ($\sigma_f = -\sigma_b$) and $P = -\sigma_f$, $D = -4\pi\sigma_f$. For an isolated crystal (D = 0) we have $E/4\pi = P/\varepsilon$ and, in the case of the parallel-plate capacitor just mentioned, we obtain $P = \sigma_b$, $E = 4\pi\sigma_b/\varepsilon$. Under laboratory conditions the piezoelectric polarization P in the direct effect is measured usually in terms of the density of the free charge σ_f, which compensates the bound charge.

$$d_{nj} = \frac{\varepsilon_{mn}^t}{4\pi} g_{nj} = e_{ni} s_{ij}^E,$$

$$e_{nj} = \frac{\varepsilon_{mn}^r}{4\pi} h_{mj} = d_{ni} c_{ij}^E,$$

$$g_{nj} = 4\pi \beta_{mn}^t d_{mj} = h_{ni} s_{ij}^D, \qquad \text{(IX.37)}$$

$$h_{nj} = 4\pi \beta_{mn}^r e_{mj} = g_{ni} c_{ij}^D,$$

$$(m, n = 1, 2, 3; \ i, j = 1, 2, \ldots, 6).$$

We can easily see that the piezoelectric coefficients e and h have the dimensions of the polarization ($cm^{-1/2} \cdot g^{1/2} \cdot sec^{-1}$ in the cgs esu system) and the coefficients d and g have the dimensions of the reciprocal of the electric field ($cm^{1/2} \cdot g^{-1/2} \cdot sec^{-1}$ in the cgs esu system).

We have stressed earlier that the values of the coefficients in the equations given so far are the adiabatic values. Adiabatic conditions (dQ = 0 or S = const) are obtained when a crystal does not lose or acquire heat during measurement. This is true of high-frequency elastic vibrations of crystals in which heat exchange between different parts of a crystal is negligible.

The other extreme case is represented by isothermal conditions (T = const). These conditions are obtained when changes take place so slowly that a crystal is always in equilibrium with its ambient medium (static measurements).

We shall not consider this point in detail but we shall simply mention that the relative difference between the adiabatic and isothermal values of the elastic constants s_{ij} is about 1%. Usually, the adiabatic constants are smaller than the isothermal values. The difference between the adiabatic and isothermal piezoelectric coefficients of linear pyroelectrics is even smaller and amounts to about 0.1%. The isothermal and adiabatic piezoelectric coefficients of materials which are piezoelectric but not pyroelectric (polar-neutral crystals) are equal. However, we must remember that, in the case of ferroelectrics (particularly near phase-transition temperatures), the difference between the isothermal and adiabatic coefficients may be considerable because of an appreciable difference between the magnitudes of the pyroelectric and electro-caloric effects. The adiabatic and isothermal values of the permittivity are equal for crystals which are not pyroelectric. In the case of linear pyroelectrics, this difference between the permittivities does not exceed a few thousandths of 1%. However, in the case of ferroelectrics near their phase transition points the difference between the adiabatic (ε^S) and isothermal (ε^T) permittivities may be considerable (see, for example, our discussion of triglycine sulfate crystals in Chap. VII).*

B. Conditions Necessary for the Determination of Piezoelectric Coefficients. It follows from our discussion that the piezoelectric properties of crystals can be described in four ways, using four coefficients d, e, g, and h (in general, we can introduce two additional constants by considering the relationship between strains r and stresses t, on the one side, and electric polarization, on the other). These coefficients are not independent: if we know one of them we can use other properties of a crystal to calculate the remaining three piezoelectric coefficients [Eq. (IX.37)]. The coefficients d, e, g, and h are not independent, and in a given measurement relating an electrical to a mechanical property (or conversely) we are not free to determine any one of the four coefficients. The difference between these coefficients is essentially that each represents specific measurement conditions. These specific conditions not only determine the properties which are being measured but also the mechanical and electrical states of a crystal.

We must stress the state of a crystal because dielectric properties are different for mechanically clamped and mechanical free crystals. Moreover, elastic properties depend on whether a crystal is electrically short-circuited or open-circuited. The actual electrical and mechanical states in the measurement of a given piezoelectric coefficient must be taken into account also in the calculation of other coefficients from the measured one. It is not sufficient to know the quantity relating a given pair of coefficients but also the conditions under which this quantity is measured [these conditions are represented by superscripts attached to ε, β, s, and c in the system of equations (IX.37)].

The mechanically clamped state means that a crystal cannot be deformed (r = 0). This state is indicated by the subscript r of

---

* Some of the differences between the adiabatic and isothermal values of various quantities are given by $s_{ijkl}^S - s_{ijkl}^T = -d_{ij}\alpha_{kl}T/c^t$; $\varepsilon_{ij}^S = \varepsilon_{ij}^T = -p_i p_j T/c^E$; $d_{ijk}^S - d_{ijk}^T = p_i^t(T/c^{t,E})\alpha_{jk}^E$, where α_{jk} is the thermal expansion coefficient; c is the specific heat; p is the pyroelectric coefficient [9].

the quantities ε and β. The mechanically free state corresponds to the absence of mechanical stresses during measurements (t = 0). It is indicated by the subscript t of the quantities ε and β. Sometimes it is convenient to consider the state of a crystal with a constant strain (r = const) or under a constant stress (t = const). However, no new information can be obtained by using constant stress or strain conditions.

The short-circuited state of a crystal (which is sometimes called the electrically free state) corresponds to experimental conditions under which the whole surface of the crystal is at the same potential (E = 0). This state is also obtained when a short-circuited pair of electrodes covers those surfaces on which the piezoelectric polarization appears in the measurement of the quantities s and c.

The open-circuited state corresponds to total isolation of a crystal. However, the state of total isolation does not always give rise to D = 0. For example, in the case of an isolated plate, the condition D = 0 is satisfied only when the polarization is directed normally to its large faces. In general, in a large isolated plate only the normal component of D and the tangential component of E remain constant far from the plate edges (Chap. VII, § 1).

The case of an isolated crystal with its piezoelectric polarization normal to its large faces is encountered quite frequently and the condition D = 0 [see, for example, Eq. (IX.18)] can be satisfied quite easily. If the condition D = 0 is satisfied, a crystal can be called electrically clamped. The electrically clamped state is also obtained when electrodes cover the large faces of a plate, provided these electrodes are not connected. This state differs from the state of total isolation only in one respect: when electrodes are present, the field in the interior of a dielectric is not due to polarization charges but to charges on the electrodes generated by electrostatic induction. The states with D = 0 and E = 0 (or, more generally, the states with D = const and E = const) are indicated by the superscripts D and E attached to the quantities s and c. The state with D = const corresponds to a charged capacitor with an air gap between the dielectric and the capacitor plates.

Let us consider again the conditions during measurements of piezoelectric coefficients, using the system of equations (IX.33) and the relationships defining these coefficients [for example,

Eq. (IX.21)]. Let us deal first with the conditions during the determination of the piezoelectric strain coefficient d. It follows from Eq. (IX.33) that

$$r_i = d_{ni}E_n,$$
$$D_m = 4\pi d_{mi}t_i. \qquad\qquad (IX.38)$$

Moreover, according to Eq. (IX.26)

$$d_{mj} = \frac{1}{4\pi}\left(\frac{\partial D_m}{\partial t_j}\right)_E = \left(\frac{\partial r_i}{\partial E_m}\right)_t. \qquad\qquad (IX.39)$$

It follows from the last two relationships that the coefficient d in the converse piezoelectric effect is determined when a crystal is in a mechanically free state (t = 0) and subjected to an electric field (E = const). In the direct piezoelectric effect a crystal is electrically short-circuited (its electrodes are connected, E = 0) and mechanical stresses on its surface are specified (r = 0). Thus, in the converse effect we measure the strain of a mechanically free crystal connected to a voltage source and in the direct effect we measure the charge which compensates the polarization [$D_m = 4\pi P_m$, see Eq. (IX.36)] under specified mechanical stresses (r = 0).

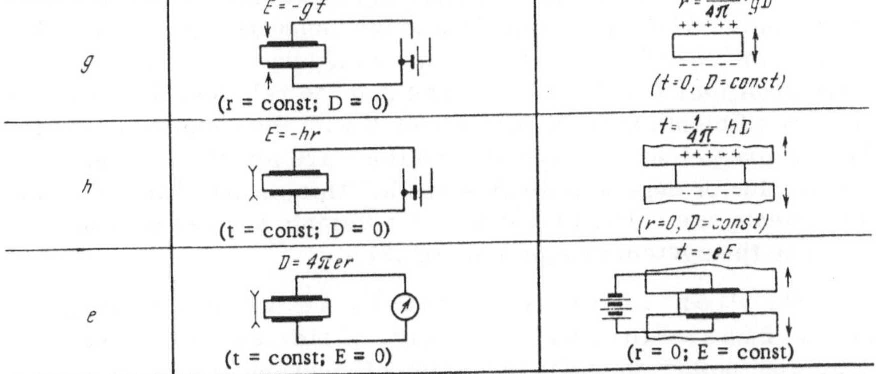

Piezoelectric coefficient	Direct effect	Converse effect
d	$D = 4\pi dt$ (r = const; E = 0)	$r = dE$ (t = 0; E = const)
g	$E = -gt$ (r = const; D = 0)	$r = \frac{1}{4\pi}\cdot gD$ (t = 0, D = const)
h	$E = -hr$ (t = const; D = 0)	$t = \frac{1}{4\pi}hD$ (r = 0, D = const)
e	$D = 4\pi er$ (t = const; E = 0)	$t = -eE$ (r = 0; E = const)

Fig. 239. Schematic representation of the conditions during the measurement of the piezoelectric coefficients d, g, h, and e.

Figure 239 shows graphically the conditions during the determination of all the piezoelectric coefficients associated with the direct and converse effects. The conditions specified in a given measurement are shown on the left-hand side of each circuit. In the case of the direct effect, the specified conditions are mechanical stresses (represented by two separate arrows pointing toward each other) and mechanical strains (two joined arrows pointing toward each other). In the converse piezoelectric effect the specified conditions are represented either by a battery (a voltage source) or a source of charge (charges are represented by the symbols + and −). It is worth mentioning here that the condition E = const in the converse piezoelectric effect can be satisfied also by an alternating field if the field frequency is lower than the frequency of natural vibrations of a crystal. An instrument used in measurements or the quantity being measured is shown on the right-hand side of each circuit. In the direct piezoelectric effect the charge is measured with a galvanometer (represented by a circle with a pointer) and the potential difference is measured with an electrometer (represented by a battery and the electrometer filament). In the converse effect two joined arrows, pointing away from each other, indicate that the strain is measured; two separate arrows, also pointing away from each other, indicate that the stress is measured (in this case, the crystal is clamped, which is also shown in the figure). We can easily see from Fig. 239 that the coefficients h and e are easiest to measure in the direct effect and the coefficient d can be measured most easily in the converse effect.

Figure 239 also shows clearly the relationships between the coefficients. Thus, for example, the piezoelectric coefficients d and g in the converse effect are determined under identical mechanical conditions (t = 0). This means that the relationships between these coefficients have factors ε and β with the superscript t. In the converse effect, electrical conditions are identical in the determination of the coefficients g and h. This means that these two coefficients are related by elastic coefficients such as s^D and c^D, etc. [see the system of equations (IX.37)].

C. Piezoelectric Crystals as Electromechanical Transducers. The ability of piezoelectric crystals to become polarized under the action of mechanical stresses and to deform under the action of an electric field makes it possible to use them as electromechanical transducers. The efficiency of such transducers depends on the difference between the permittivities of the mechanically clamped and mechanically free states of a crystal and on the difference between the elastic coefficients of the electrically short-circuited and open-circuited states.

In the case of a mechanically clamped crystal, the energy of an electric field in the measurement of ε is used entirely to set up polarization. If a crystal is free, some of this energy is transformed into the energy of mechanical vibrations. In measurements of the elastic constants s of an open-circuited crystal ($D = 0$) some mechanical energy is transformed into the energy of an electric field. Conversely, if a crystal is short-circuited no such energy transformation takes place. A qualitative relationship between the values of ε^r, ε^t, s^E, and s^D can be found on the basis of the following considerations. The polarization of a free crystal is due to the direct action of the field on charges as well as due to the deformation of the crystal (the converse piezoelectric effect) and the signs of these two contributions to the polarization are opposite. It follows that $\varepsilon^r < \varepsilon^t$.

In measurements of the elastic coefficients (i.e., elastic constants and elastic moduli) of an open-circuited crystal the piezoelectric polarization charge (the depolarization field) deforms the crystal because of the converse effect along a direction opposite to the direction of the initial stress. Under these conditions the crystal becomes effectively more rigid. It follows that $s^D < s^E$ (an open-circuited crystal is less compliant).

The efficiency of electromechanical conversion of energy can be represented by the ratio of the energy which appears in the mechanical form to the total electrical energy supplied by a battery. This ratio is represented by a symbol k^2; k is the electromechanical coupling coefficient. We can easily see that k^2 can be expressed in terms of the coefficients s^E, s^D, ε^t, and ε^r:

$$\frac{s^E - s^D}{s^E} = \frac{\varepsilon^t - \varepsilon^r}{\varepsilon^t} = k^2. \qquad (IX.40)$$

Let us now consider the conversion associated with an i-th component of the electric field or induction and a j-th component of strain r or stress t. We can easily obtain the following expressions for the electromechanical coupling coefficient in terms of the piezoelectric coefficients:

$$k^2 = \frac{d_{ij}^2}{s_{ii}^t s_{jj}^E} = \frac{h_{ij}^2}{\beta_{ii}^r c_{jj}^D},$$

$$k^2 = \frac{e_{ij}^2}{s_{ii}^r c_{jj}^E} = \frac{g_{ij}^2}{\beta_{ii}^t s_{jj}^D}. \tag{IX.41}$$

It follows directly from the last two expressions that the electro-mechanical coupling coefficient is proportional to the piezoelectric coefficients.

For the majority of linear piezoelectrics the electromechanical coupling coefficient is of the order of 10%, which corresponds to a difference of about 1% between the values of s^D and s^E or between ε^t and ε^r; such differences are not of practical importance. However, the small value of k does not mean that a given piezoelectric is a poor material: in some applications of piezoelectric crystals the energy relationships are of no importance. The value of k of ferroelectrics may reach 90% because the difference between the values of s (or ε) may reach 8%.

The various piezoelectric coefficients discussed so far represent the operation of piezoelectric transducers working under different conditions. Thus, it is evident from Fig. 239 that if our aim is to obtain high voltages by the application of mechanical stresses we must use crystals which have the largest possible coefficients g. These requirements are encountered, for example, in piezoelectric detectors of sound and ultrasound. If our purpose is to produce strong strains in a crystal by the application of an electric field we must look for crystals with the highest values of the coefficients d. Such requirements are encountered, for example, in the construction of efficient sources of sound and ultrasound. If our aim is to convert efficiently mechanical strains by the application of electric fields to piezoelectric crystals we must use materials with the highest values of the coefficients h or e (such applications include a piezoelectric seismograph and a piezoelectric acoustic pickup).

Thus, a piezoelectric transducer is characterized by different piezoelectric coefficients under different working conditions. It must be remembered that a large value of one of the piezoelectric coefficients is not necessarily associated (in the same crystal) with large values of all the other coefficients. The efficiency of a piezoelectric crystal as a transducer is a function of many of its mechanical and electrical properties.

We must stress once again that the basic consideration in the applications of piezoelectric crystals is not just the conversion efficiency. It is worth mentioning that the direct transformation of mechanical into electrical energy by means of piezoelectric crystals as such is not used at all in practical applications. In other applications the important aspects are (apart from the energy relationships) such properties of a piezoelectric crystal as its mechanical and electric strength, resistance to the effects of moisture, stability of dielectric properties, weak temperature dependences of the properties, and so on. Nevertheless, other conditions being equal, it is desirable to use — in a given device — those crystals which have the largest values of the relevant piezoelectric coefficients.

§3. Piezoelectric Properties of Some Linear Dielectrics

There are now about 1000 substances known to possess piezoelectric properties. The majority of them are linear piezoelectrics. Detailed investigations of piezoelectric properties have been carried out for only a few tens of crystals. We shall describe the piezoelectric properties of the most important substances. To give a comprehensive description of a given material we shall discuss not only its piezoelectric but also its elastic and dielectric properties.

A. Quartz. Some information on the structure and crystal symmetry of quartz is given in Chap. I.

The low-temperature modification of SiO_2, known as β quartz, was one of the first crystals in which Pierre and Jacques Curie discovered piezoelectric properties in 1880. The point group of β quartz (from now on we shall refer to it simply as quartz) is 3:2, its class is trigonal trapezohedral, and the system is trigonal. An ideal quartz polyhedron (with faces encountered in naturally occurring samples) is shown in Fig. 240 in its two enantiomorphous (left- and right-handed) modifications. The physical system of coordinates is also given in Fig. 240.

The tensor equations describing the direct and converse piezoelectric effects (using one of the piezoelectric strain coeffi-

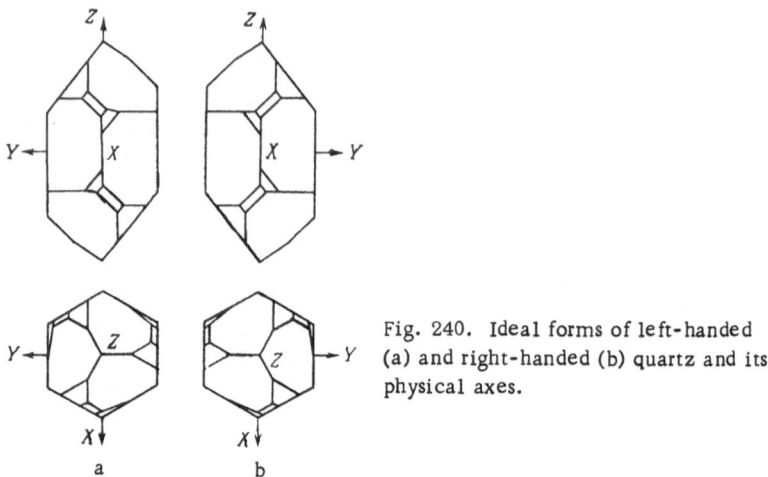

Fig. 240. Ideal forms of left-handed (a) and right-handed (b) quartz and its physical axes.

a   b

cients d_{ij}) for class 3:2 are

	t_1	t_2	t_3	t_4	t_5	t_6
P_1	d_{11}	$-d_{11}$	0	d_{14}	0	0
P_2	0	0	0	0	$-d_{14}$	$-2d_{11}$
P_3	0	0	0	0	0	0

$$\text{(IX.42)}$$

for the direct effect, and

	E_1	F_2	E_3
r_1	d_{11}	0	0
r_2	$-d_{11}$	0	0
r_3	0	0	0
r_4	d_{14}	0	0
r_5	0	$-d_{14}$	0
6	0	$-2d_{11}$	0

$$\text{(IX.43)}$$

for the converse effect.

The usual form of the equations for the direct piezoelectric effect is:

$$P_1 = d_{11}t_1 - d_{11}t_2 + d_{14}t_4,$$
$$P_2 = -d_{14}t_5 - 2d_{11}t_6;$$

$$\text{(IX.44)}$$

the equations for the converse effect are:

$$r_1 = d_{11}E_1,$$
$$r_2 = - d_{11}E_1,$$
$$r_4 = d_{14}E_1,$$
$$r_5 = - d_{14}E_2,$$
$$r_6 = - 2d_{11}E_2.$$

(IX.45)

The most characteristic property of the piezoelectric effect in quartz is that it does not exhibit piezoelectric properties along the Z axis (optic axis). The simplest piezoelectric cuts of quartz are the X cut (or the Curie cut) and the Y cut (as usual, the X axis is denoted by the subscript 1, the Y axis by the subscript 2, and the Z axis by the subscript 3). X-cut plates are usually employed to excite the longitudinal piezoelectric effect (Fig. 241a) and Y-cut plates are used to excite the transverse effect (Fig. 241b). The longitudinal piezoelectic effect for the X cut is

$$P_1 = d_{11}t_1,$$

(IX.46)

and the transverse effect for the Y cut is

$$P_1 = - d_{11}t_2.$$

(IX.47)

The piezoelectric polarization due to shear stresses (Fig. 241c) is described by the piezoelectric strain coefficient d_{14}.

The most reliable values of the piezoelectric strain coefficients of quartz, expressed in cgs esu, are: $d_{11} = -6.76 \times 10^{-8}$ and $d_{14} = +2.56 \times 10^{-8}$.

In order to obtain an idea of the magnitude of the converse piezoelectric effect in quartz we must mention that a voltage of

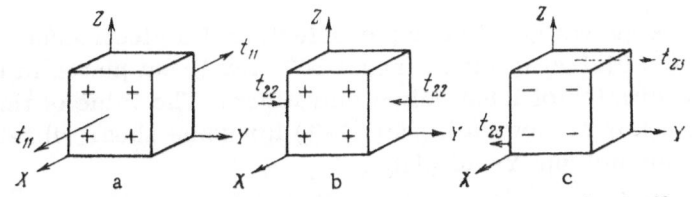

Fig. 241. Direct piezoelectric effect in quartz: a) longitudinal effect; b) transverse effect; c) effect produced by shear.

1000 V, applied along the X axis to an X-cut plate 1 cm thick, in-creases its thickness by 21 Å [Eq. (IX.45)]. In the direct piezo-electric effect the same plate produces a potential difference of 60 V along the X axis when a stress of 1 kgf/cm^2 is applied along this axis.

The adiabatic and isothermal values of the elastic moduli c and the elastic constants s of quartz are practically identical. The room-temperature values of these elastic coefficients, measured in a fixed field, are as follows (in cgs esu):

$$s_{11}^E = 127.9 \cdot 10^{-14}, \qquad\qquad c_{11}^E = 86.05 \cdot 10^{10},$$

$$s_{12}^E = -15.35 \cdot 10^{-14}, \qquad\qquad c_{12}^E = 4.85 \cdot 10^{10},$$

$$s_{13}^E = -11.0 \cdot 10^{-14}, \qquad\qquad c_{13}^E = 10.45 \cdot 10^{10},$$

$$s_{14}^E = -44.6 \cdot 10^{-14}, \qquad\qquad c_{14}^E = 18.25 \cdot 10^{10},$$

$$s_{33}^E = 95.6 \cdot 10^{-14}, \qquad\qquad c_{33}^E = 107.1 \cdot 10^{10},$$

$$s_{44}^E = 197.8 \cdot 10^{-14}, \qquad\qquad c_{44}^E = 58.65 \cdot 10^{10},$$

$$s_{66}^E = 2\,(s_{11}^E - s_{12}^E) = 286.5 \cdot 10^{-14}, \qquad c_{66}^E = \frac{c_{11}^E - c_{12}^E}{2} = 40.5 \cdot 10^{10}.$$

Quartz is optically and electrically uniaxial and has two in-dependent permittivities $\varepsilon_1 = \varepsilon_2$ and ε_3. The values of these per-mittivities, measured under a constant mechanical stress, are: $\varepsilon_1^t = 4.58$ and $\varepsilon_3^t = 4.70$.

As mentioned in Chap. VIII, the electrical conductivity of quartz depends strongly on the presence of impurities and, therefore, it varies from sample to sample. The only common property of different samples seems to be the higher electrical conductivity along the optic axis than at right-angles to this axis. In rough calculations the room-temperature resistivity of quartz along the optic axis (Z axis) is usually taken to be 10^{14} $\Omega \cdot$ cm and at right-angles to this axis the resistivity is assumed to be 2 × 10^{16} $\Omega \cdot$ cm.

Since quartz is a linear piezoelectric, its electromechanical coupling coefficient is small (about 10%) for X-cut plates in the case of excitation of longitudinal vibrations. The value of this coefficient k is somewhat higher (14%) for some shear vibrations in Y-cut and oblique Y-cut plates.

The wide use of quartz in various (including piezoelectric) devices is due to its many valuable properties. Thus, the hardness of quartz on the Moh's scale of hardness is 7; quartz does not dissolve in water, resists various acids, melts at 1700°C, and has small thermal expansion coefficients ($\alpha_1 = 8 \times 10^{-6}$ deg^{-1}, $\alpha_3 = 13.4 \times 10^{-6}$ deg^{-1}). The specific gravity of quartz is 2.65 g/cm^3. The excellent insulating properties of quartz have been mentioned earlier.

Optically, quartz is a uniaxial positive crystal ($n_1 = 1.55336$, $n_3 = 1.54425$ for $\lambda = 589.3$ mμ), which is transparent in the ultraviolet and (partly) in the infrared range. Quartz is optically active: in the visible part of the spectrum the angle of rotation of the plane of polarization about the Z axis is $\sim 25'$ per 1 mm of thickness.

The valuable properties of quartz include the weak temperature dependences of many of its properties (this is true between very low temperatures and the $\beta \rightarrow \alpha$ transition point at $+573°C$). The thermal stability of the properties of quartz is responsible for its wide use, particularly in those cases where high precision is required and the parameters of radio circuits must remain constant.

The piezoelectric properties of quartz are used mainly in the frequency stabilization of rf oscillators, in highly selective filters, and in high-frequency electromechanical transducers used to excite or detect mechanical vibrations in gases, liquids, and solids. Quartz is used also in pressure measurements.

A piezoelectric quartz element in the form of a resonator (an element with a holder, protective cap, etc.) has definite equivalent parameters (a capacitance C, an inductance L, and a resistance R) which represent its properties as an oscillatory circuit. A valuable characteristic of piezoelectric quartz resonators is their extremely high Q factor (the ratio of the amplitude of steady-state forced oscillations at resonance to the amplitude of forced oscillations far from resonance). For quartz resonators the value of this factor is 10^5–10^6.

The high value of the Q factor of a quartz oscillator ensures favorable conditions of its operation at the resonance frequency, which is the frequency generated by such an oscillator.

Circuits of oscillators with stabilizing quartz resonators (usually called quartz oscillators) do not differ basically from conventional oscillatory circuits. Quartz oscillators are usually em-

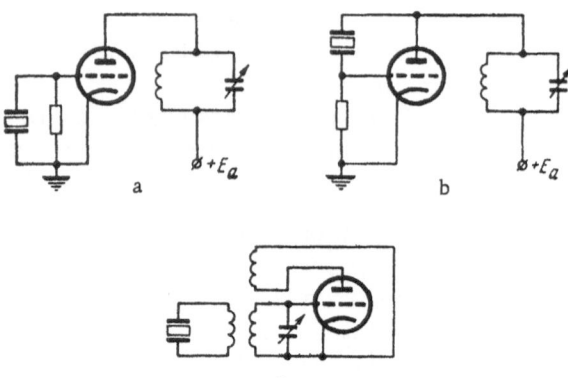

Fig. 242. Quartz circuits: a, b) oscillator circuits; c)
frequency pulling circuit.

ployed in two types of circuit: an oscillatory circuit (Figs. 242a
and 242b) or a frequency pulling circuit (Fig. 242c). In the circuit of Fig. 242a a quartz crystal connected between the grid and
the cathode of an electron tube acts as a conventional oscillatory
circuit with a high value of Q. The phase and amplitude of the
negative feedback (via the anode–grid capacitance of the tube), required for the self-excitation of such an oscillator, are selected
by tuning an additional oscillatory circuit connected to the anode
of the tube. In the circuit of Fig. 242b a quartz crystal is inserted
between the grid and anode of an electron tube. An oscillatory circuit connected to the anode is again used to control negative feedback via the grid–cathode capacitance. In the frequency pulling
circuit (Fig. 242c) quartz is coupled inductively to the oscillatory circuit. The stabilizing effect of the quartz circuit is based
on the pulling effect in which the secondary high-Q quartz circuit
imposes its frequency on the primary circuit when the latter de-
parts slightly from its resonance frequency. The frequency pulling circuits are less stable than other quartz circuits and are not
used extensively.

The frequency stability of a quartz oscillator can be represented by the ratio $\Delta f/f$ (Δf is the frequency drift from its
nominal value) which is usually 10^{-3}-$10^{-5}\%$. In some special circuits, such as time standards in the form of quartz clocks (which
are kept under special conditions ensuring minimum variations of
temperature and humidity as well as absence of vibrations), the

relative stability is 10^{-10} per day which corresponds to a "drift" of such a clock by 1 sec in 70 years. Quartz oscillators can operate at frequencies from a few kilocycles to tens or even hundreds of megacycles.

A piezoelectric quartz resonator can also be used as an electric filter passing only those frequencies which are close to the resonance frequency of its circuit. The high value of the Q factor and the selectivity of quartz can be used to construct quartz filters with a narrow pass band in the frequency range from a few kilocycles to hundreds of kilocycles. Such quartz filters are used in multichannel telephony and telegraphy for the separation of signals transmitted along different channels, as well as in radio receivers where they are employed to increase the selectivity. The relative width of the pass band of a quartz filter depends on the actual circuit, but it is generally small, amounting to 0.4-0.8% of the resonance frequency. A wider pass band, required in many cases, can be achieved by connecting induction coils in series with quartz resonators. In such devices the relative width of the pass

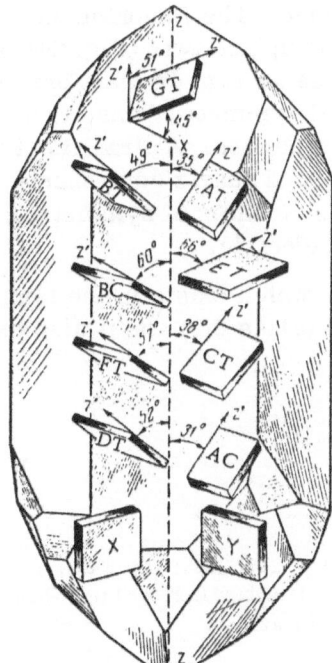

Fig. 243. Various cuts of piezoelectric quartz.

570

CHAPTER IX

band may reach 12% of the resonance frequency. Recent years have seen the introduction of piezoelectric filters made of ethylene diamine tartrate, potassium tartrate, and piezoelectric ceramics.

The high stability of quartz oscillators depends to a great extent on the temperature dependence of the frequency of a quartz resonator. Studies of the nature of the thermal expansion of quartz show that it can be cut into plates for which the temperature dependence of the frequency (represented by the temperature coefficient of the frequency) is very weak or even absent in a wide range of temperatures (these are known as cuts of zero temperature coefficient). The stability of a resonator is governed, apart from the small value of the temperature coefficient of the frequency, by other factors: the ability to obtain the required frequency using a plate cut in a particular way, the ability to excite a particular vibration, the nature of clamping of the plate, and so on. In view of this variety of requirements, a large number of cuts is used, some of which are shown in Fig. 243.

Concluding our discussion of the piezoelectric properties and applications of quartz, we must mention that naturally occurring quartz crystals are frequently twinned. The Dauphiné and Brazilian quartz twins are widely known. A Dauphiné twin (Fig. 244a) consists of identical crystals with the same sign of enantiomorphism. Such an ideal twin has the hexagonal symmetry (class 6:2). Dauphiné twins exhibit piezoelectric properties. A Brazilian twin (Fig. 244b) consists of right- and left-handed quartz components. Such an ideal twin has an inversion center (center of symmetry), belongs to class $\bar{6}\cdot m$, and is not piezoelectric.

The α modification of quartz, which exists in the temperature range 573–870°C, is hexagonal (class 6:2) and it also exhibits piezoelectric properties.

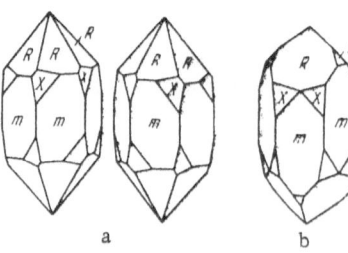

Fig. 244. Ideal quartz twins:
a) Dauphiné (left- and right-handed);
b) Brazilian.

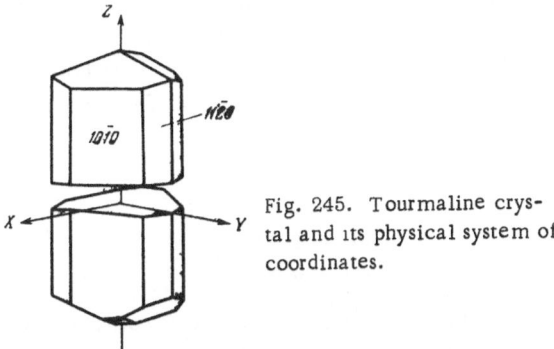

Fig. 245. Tourmaline crystal and its physical system of coordinates.

B. Tourmaline. Tourmaline crystals (for structure, see Chap. I) are not widely used and are interesting primarily because they exhibit piezoelectric polarization under hydrostatic pressure and because they have pyroelectric properties.

Tourmaline crystals belong to the trigonal system of symmetry (ditrigonal pyramidal class 3·m); the threefold axis of these crystals is unique and polar. The physical system of coordinates of tourmaline is shown in Fig. 245. Like quartz, tourmaline is a hard mineral (7–7.5 on Moh's scale), it is insoluble in water, and its specific gravity is 2.90–3.25 g/cm³.

The tensor of the elastic coefficients of tourmaline is of the same type as that of quartz. The most reliable room–temperature values of the elastic moduli c and the elastic constants s (in cgs esu) are as follows:

$$c_{11} = 272 \cdot 10^{10}, \qquad s_{11} = 0.385 \cdot 10^{-12},$$
$$c_{12} = 40 \cdot 10^{10}, \qquad s_{12} = 0.048 \cdot 10^{-12},$$
$$c_{13} = 35 \cdot 10^{10}, \qquad s_{13} = 0.071 \cdot 10^{-12},$$
$$c_{33} = 165 \cdot 10^{10}, \qquad s_{33} = 0.636 \cdot 10^{-12},$$
$$c_{44} = 65 \cdot 10^{10}, \qquad s_{44} = 154 \cdot 10^{-12},$$
$$c_{14} = -6.8 \cdot 10^{10}, \qquad s_{14} = 0.045 \cdot 10^{-12}.$$

The matrix of the piezoelectric strain coefficients of tourmaline (Table 46) is

$$\begin{pmatrix} 0 & 0 & 0 & 0 & d_{15} & -d_{22} \\ -d_{22} & d_{22} & 0 & d_{15} & 0 & 0 \\ d_{31} & d_{31} & d_{33} & 0 & 0 & 0 \end{pmatrix}. \qquad \text{(IX.48)}$$

It is evident from Eq. (IX.48) that tourmaline has four independent piezoelectric strain coefficients and that the piezoelectric effect is observed along all three physical axes. The room-temperature values of the piezoelectric strain coefficients of tourmaline (in cgs esu) are:

$$d_{15} = 10.9 \cdot 10^{-8},$$
$$d_{22} = 1.0 \cdot 10^{-8},$$
$$d_{31} = 1.03 \cdot 10^{-8},$$
$$d_{33} = 5.5 \cdot 10^{-8}.$$

The piezoelectric strain coefficient for hydrostatic compression, calculated using the values just quoted, is

$$d_{\text{hydr}} = (d_{31} + d_{32} + d_{33}) = 7.56 \cdot 10^{-8} \text{ cgs esu}.$$

The piezoelectric polarization of Z-cut plates subjected to hydrostatic pressure is represented by this coefficient.

Tourmaline is an electrically uniaxial crystal; the permittivities measured under a constant stress are:

$$\varepsilon_{11}^{t} = \varepsilon_{22}^{t} = 8.2,$$
$$\varepsilon_{33}^{t} = 7.5.$$

The refractive indices of tourmaline are:

$$n_1 = n_2 = 1.6193,$$
$$n_3 = 1.6366.$$

Tourmaline crystals have been used in piezoelectric resonators (mainly as Z-cut plates with thickness expansion-compression vibrations), but tourmaline resonators are not widely used. The pyroelectric properties of tourmaline and its polarization under hydrostatic pressure have been used in the development of some precision measuring instruments.

C. Potassium Tartrate. Some salts of tartaric acid, $C_4H_4O_6$, have remarkable electrical properties. The best known is Rochelle salt, whose piezoelectric properties will be discussed in the next section. Here we shall consider only the piezoelectric properties of potassium tartrate, $K_2C_4H_4O_6 \cdot \frac{1}{2} H_2O$, which is a linear dielectric. Crystals of the tartrate group include also ethylene diamine tartrate, $C_6H_{14}N_2O_6$, which is a piezoelectric and is fre-

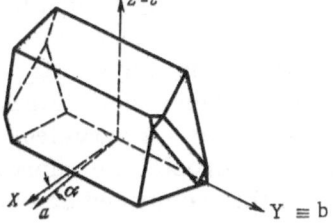

Fig. 246. Potassium tartrate crystal with its crystallographic and physical systems of coordinates ($\alpha = 0°51'$).

quently used in resonators and filters. However, we shall not discuss ethylene diamine tartrate because it has now been largely replaced by potassium tartrate. The potassium compound has several advantages over ethylene diamine tartrate; in particular, it is easier to grow potassium tartrate (both tartrates are used in the synthetic form).

Crystals of potassium tartrate are grown from aqueous solutions and their specific gravity is 1.988 g/cm³. Under normal humidity conditions (50-60%) they are quite stable and have good dielectric properties.

Potassium tartrate belongs to the monoclinic system (dihedral axial class 2). In the crystallographic system of axes, axis 2 is taken to be the b axis and the a and c axes lie in the planes of the best developed faces. The unit-cell parameters are $a = 20.101$ Å, b = 5.049 Å, c = 15.490 Å, $\beta = 90°51'$. The physical axes are shown in Fig. 246.

TABLE 51. Values of the Principal Parameters of Potassium Tartrate Crystals (cgs esu)

Piezoelectric coefficients d, 10^{-8}	Elastic constants s, 10^{-12}		Permittivity ε	Thermal expansion coefficients α, 10^{-6} deg^{-1}
$d_{14} = -25.0$	$s_{11} = +2.24$	$s_{33} = +3.86$	$\varepsilon_{11} = 6.44$	$\alpha_{11} = +12.0$
$d_{16} = +6.5$	$s_{12} = -0.08$	$s_{35} = +0.90$	$\varepsilon_{22} = 5.80$	$\alpha_{22} = +44.8$
$d_{21} = -2.2$	$s_{13} = -1.64$	$s_{44} = +11.90$	$\varepsilon_{33} = 6.49$	$\alpha_{33} = +32.0$
$d_{22} = 8.5$	$s_{15} = -0.64$	$s_{46} = +0.57$	$\varepsilon_{13} = 0.005$	$\alpha_{13} = -12.0$
$d_{23} = -10.4$	$s_{22} = +3.37$	$s_{55} = +8.15$		
$d_{25} = -22.5$	$s_{23} = -1.05$	$s_{66} = +10.41$		
$d_{34} = +29.4$	$s_{25} = -0.57$			
$d_{36} = -66.0$				

The values of the principal parameters of potassium tartrate (in cgs esu) are listed in Table 51. Examination of this table shows that the shear piezoelectric strain coefficients (d_{14}, d_{25}, and d_{36}) are large compared with the corresponding coefficients of quartz and tourmaline. Potassium tartrate is a pyroelectric and exhibits the piezoelectric effect under hydrostatic compression. The piezoelectric coefficient d_{hydr} of this compound can be found from the matrix of the piezoelectric coefficients of crystals of class 2

$$\begin{pmatrix} 0 & 0 & 0 & d_{14} & 0 & d_{16} \\ d_{21} & d_{22} & d_{23} & 0 & d_{25} & 0 \\ 0 & 0 & 0 & d_{34} & 0 & d_{36} \end{pmatrix}, \qquad \text{(IX.49)}$$

using the expression $d_{hydr} = d_{21} + d_{22} + d_{23} = -4.1 \times 10^{-8}$ cgs esu, which is about half the corresponding hydrostatic coefficient of tourmaline. The relatively high values of the electromechanical coupling coefficients of ethylene diamine and potassium tartrates (20–25%), can be used to construct band-pass filters of sufficient band width without the use of induction coils (necessary in the case of quartz), which makes them cheaper and lighter.

D. Lithium Sulfate. Crystals of lithium sulfate monohydrate, $LiSO_4 \cdot H_2O$, are usually prepared from aqueous solutions or from melts. These crystals are monoclinic and, like those of potassium tartrate, they belong to the dihedral axial class 2. The unit-cell parameters are: $a = 8.18$ Å, $b = 4.87$ Å, $c = 5.45$ Å, $\beta = 107°18'$. The crystallographic and physical systems of axes are shown in Fig. 247 (the b axis coincides with axis 2).

The principal elastic, piezoelectric, and dielectric parameters of lithium sulfate are listed in Table 52. A remarkable property of lithium sulfate is its very strong piezoelectric polarization under hydrostatic compression. It follows from Eq. (IX.49) and

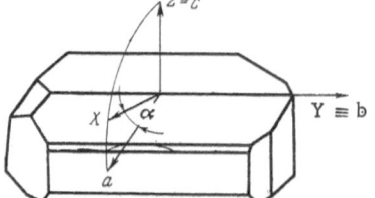

Fig. 247. Lithium sulfate crystal with its crystallographic and physical systems of coordinates ($\alpha = 17°18'$).

TABLE 52. Values of the Principal Parameters of Lithium Sulfate Crystals (cgs esu)

Piezoelectric coefficients d, 10^{-8}	Elastic constants s, 10^{-12}		Permittivity ε
$d_{14} = +14.0$	$s_{11}^E = 2.39$	$s_{23}^E = -0.36$	$\varepsilon_{11}^t = 5.6$
$d_{16} = -12.5$	$s_{22}^E = 2.13$	$s_{15}^E = +0.71$	$\varepsilon_{22}^t = 6.5$
$d_{21} = +11.6$	$s_{33}^E = 2.31$	$s_{25}^E = -1.20$	$\varepsilon_{33}^t = 10.2$
$d_{22} = -45.0$	$s_{44}^E = 3.69$	$s_{35}^E = +0.05$	$\varepsilon_{13}^t = 0.07$
$d_{23} = -5.5$	$s_{55}^E = 4.1$	$s_{46}^E = -0.41$	
$d_{25} = +16.5$	$s_{66}^E = 7.40$		
$d_{34} = -26.4$	$s_{12}^E = -0.95$		
$d_{36} = +10.0$	$s_{13}^E = -0.5$		

Table 52 that d_{hydr} of this crystal is:

$$d_{hydr} = -38.9 \cdot 10^{-8} \text{ cgs esu.}$$

It is also worth noting the high value of the piezoelectric strain coefficient d_{22}, which is almost an order of magnitude higher than the coefficient d_{11} of quartz.

E. Resorcinol. A strong piezoelectric effect is exhibited by crystals of resorcinol, $C_6H_4(OH)_2$, grown from aqueous solutions [5]. The specific gravity of these crystals is 1.272 g/cm^3 and their melting point is about 100°C. Resorcinol crystals are orthorhombic (rhombic pyramidal class 2·m) and they exhibit piezoelectric and pyroelectric properties. The physical system of axes of resorcinol is shown in Fig. 248 (axis 2 is used as the X_3 axis).

The principal elastic, piezoelectric, and dielectric parameters of resorcinol are listed in Table 53.

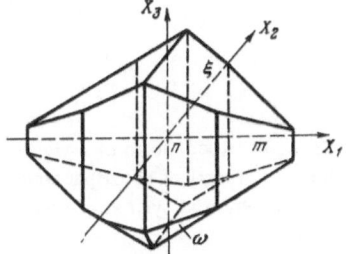

Fig. 248. Physical axes of resorcinol.

TABLE 53. Values of the Principal Parameters
of Resorcinol Crystals (cgs esu)

Piezoelectric coefficients d, 10^{-8}	Elastic moduli c, 10^{10}	Elastic constants s, 10^{-12}	Permittivity ε
$d_{15} = 53.9$	$c_{11} = 10.3$	$s_{11} = 19.0$	$\varepsilon_{11} = 3.51$
	$c_{22} = 14.4$	$s_{22} = 10.6$	
$d_{24} = 55.3$	$c_{33} = 12.9$	$s_{33} = 15.0$	$\varepsilon_{22} = 4.14$
$d_{31} = -12.4$	$c_{44} = 3.3$	$s_{44} = 30.7$	$\varepsilon_{33} = 3.54$
	$c_{55} = 4.4$	$s_{55} = 23.0$	
$d_{32} = -12.8$	$c_{66} = 4.0$	$s_{66} = 25.0$	
$d_{33} = 16.8$	$c_{12} = 6.2$	$s_{12} = -4.0$	
	$c_{13} = 7.4$	$s_{13} = -3.4$	
	$c_{23} = 6.9$	$s_{23} = -8.8$	

F. Benzophenone. The stable orthorhombic modification of benzophenone, $(C_6H_5)_2 \cdot CO$, has high values of the piezoelectric coefficients g, which are among the highest known (this compound has high values of the piezoelectric strain coefficients d and low values of the permittivity).

Benzophenone crystals are grown from acetone and carbon tetrachloride solutions. Their specific gravity is 1.219 g/cm³ and the melting point is 47°C. These crystals are orthorhombic and belong to the rhombic tetrahedral class 2:2. The physical system of coordinates is shown in Fig. 249.

The principal elastic, piezoelectric, and dielectric parameters of benzophenone crystals are listed in Table 54 [15].

G. Cancrinite. Among natural piezoelectric minerals it is worth mentioning, in addition to quartz and tourmaline, the

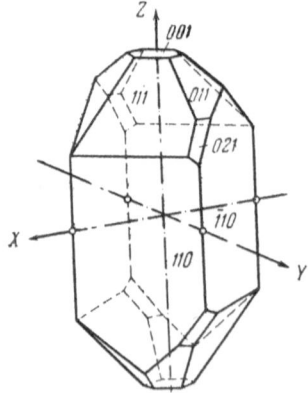

Fig. 249. Ideal form of a right-handed crystal of benzophenone and its physical system of coordinates.

TABLE 54. Values of the Principal Parameters of Benzophenone Crystals (cgs esu)

Piezoelectric coefficients d, 10^{-8}	Elastic moduli c, 10^{10}	Dielectric parameters
$d_{14} = 37.0$	$c_{11} = 10.70$	$\varepsilon_{11} = 4.0$
$d_{25} = 6.0$	$c_{22} = 10.00$	$\varepsilon_{22} = 4.1$
$d_{36} = 61.0$	$c_{33} = 7.10$	$\varepsilon_{33} = 3.7$
	$c_{44} = 2.03$	$\tan\delta = 6\cdot10^{-4}$
	$c_{55} = 1.55$	$k \cong 10-15\%$
	$c_{68} = 3.53$	
	$c_{23} = 3.21$	
	$c_{31} = 1.69$	
	$c_{12} = 5.50$	

crystals of cancrinite, which is a sodium–calcium alumosilicate, $(Na_2Ca)_4 \cdot [AlSiO_4]_6 \cdot CO_3(H_2O)_{0-3}$. The hardness of cancrinite on Moh's scale is 5-5.5; its specific gravity is 2.42-2.48 g/cm³. Cancrinite belongs to the dihexagonal pyramidal class 6·m of the hexagonal system (the attribution of its structure to this class is more convincing than that to class 2, as suggested by some workers). Some of the physical parameters of natural cancrinite crystals are given in Table 55 [6].

H. Other Crystals. We have described the piezoelectric properties of crystals belonging to crystallographic systems of low and moderately high symmetry but we have not described the piezoelectric properties of even a single cubic material. This is because the number of cubic piezoelectrics is limited and little is known about their properties.

It is worth mentioning the piezoelectric properties of two cubic crystals: sodium chlorate, $NaClO_3$ (class 3/2) and the sphalerite modification of ZnS (class $3/\overline{4}$), whose structure is described in Chap. I.

TABLE 55. Values of the Principal Parameters of Cancrinite Crystals (cgs esu)

Piezoelectric coefficients d, 10^{-8}	Elastic constants s, 10^{-12}	Permittivity ε
$d_{33} = 13.0$	$s_{11} = 2$	$\varepsilon_{11} = 9.5$
$d_{15} = 27$	$s_{12} \cong 0$	$\varepsilon_{33} = 11.2$
$d_{31} = 2.0$	$s_{13} = -0.3$	
	$s_{33} = 1.3$	
	$s_{44} = 4.2$	
	$s_{66} = 3.5$	

Each of these two crystals has one piezoelectric strain co-efficient d_{14}, three elastic coefficients, and one value of the permittivity ε.

The parameters of sodium chlorate (in cgs esu) are:

$$d_{14} = 6.1 \cdot 10^{-8},$$
$$\varepsilon^t = 5.76,$$
$$s_{11} = 2.335 \cdot 10^{-12},$$
$$(2s_{12} + s_{44}) = 7.51 \cdot 10^{-12},$$
$$s_{12} = -0.515 \cdot 10^{-12},$$
$$k \cong 3\%.$$

The parameters of the sphalerite modification of ZnS (in cgs esu) are:

$$d_{14} = -9.7 \cdot 10^{-8},$$
$$c_{11} = 94.2 \cdot 10^{10},$$
$$c_{23} = 56.8 \cdot 10^{10},$$
$$c_{44} = 43.6 \cdot 10^{10}.$$

The value of the coefficient $e_{14} = d_{14}c_{44}$, calculated from d_{14} and c_{44} given above, is -4.2×10^4 cgs esu. This piezoelectric coefficient was first estimated by Born on the basis of the dynamic theory of lattices: Born's value of e_{14} is -2.3×10^5 cgs esu, which does not agree with experiment.

§4. Piezoelectric Properties of Ferroelectrics

All ferroelectrics exhibit piezoelectric properties at least in the range of existence of their spontaneous polarization. This follows from the fact that in the ferroelectric modification these crystals are polar (in the single-domain state); if a crystal is split into domains the individual domains are polar. It is well known that crystals of all polar (pyroelectric) classes exhibit piezoelectric properties.

A macroscopic examination of a complete crystal split into domains shows that such a crystal may or may not exhibit piezoelectric properties. All depends on whether a given crystal has piezoelectric properties in the paraelectric modification before the ferroelectric phase transition, i.e., whether or not it has an inversion center (center of symmetry) in the paraelectric modification. If a crystal does not have an inversion center, i.e., if it

is piezoelectric before the ferroelectric phase transition, we find
that, according to the principle which states that a crystal reverts to its original symmetry group after splitting into domains
(Chap. IV), such a crystal has piezoelectric properties also after
the phase transition even if it is split into domains. Typical examples of phase transitions of this type are those associated with
the appearance of the spontaneous polarization in potassium dihydrogen phosphate and Rochelle salt. Crystals which are centrosymmetric in the paraelectric phase have a centrosymmetric configuration of domains and show no piezoelectric properties in the
multidomain state. Such phase transitions are exhibited, for example, by barium titanate and triglycine sulfate.

In general, the presence of domains in ferroelectrics gives
rise to some special properties of the piezoelectric effect. Thus,
the converse effect in ferroelectrics is associated not only with
the usual strains (or the appearance of stresses) but also with domain switching. In some crystals the direct effect is also associated with domain switching.

The piezoelectric properties of ferroelectrics are usually
stronger than those of linear piezoelectrics. This is especially
true near the phase transition point. Anomalies in the piezoelectric properties of ferroelectrics near the phase transition point
are another special property of the piezoelectric effect in these
materials, which distinguishes them from linear dielectrics.

The appearance of the spontaneous polarization gives rise
to anomalies of various properties of ferroelectrics at the phase
transition point (they include elastic, dielectric, optical, and other
properties). However, we cannot be as definite about anomalies
of the piezoelectric properties. All depends on which of the piezoelectric coefficients represent the piezoelectric properties of a
crystal near its phase transition: d and e or g and h. The piezoelectric coefficients d and e change very strongly in the phase
transition region, whereas the coefficients g and h remain practically constant.* The cause of the different behavior of these
piezoelectric coefficient pairs follows from the conditions under

_____
*The coefficients g and h of ferroelectrics which exhibit the piezoelectric effect in
the paraelectric and ferroelectric modifications are approximately equal for both
modifications. In the case of ferroelectrics which are centrosymmetric in the paraelectric modification, these coefficients suddenly assume finite values and do not vary
greatly near the phase transition temperature.

which they are defined (Fig. 239). The coefficients g and h are defined so that the electrical properties are governed by the mechanical properties (and conversely) in a "pure" form, without any influence of secondary phenomena. On the other hand, the definition of the coefficients d and e postulates the constancy of E. It thus follows that a change in the permittivity (which is the most characteristic property of a ferroelectric phase transition) requires a change in the charge on the electrodes of a capacitor and this change is not related to the piezoelectric effect.

The different behavior of different piezoelectric coefficients can be seen easily in the spontaneous piezoelectric effect, i.e., the piezoelectric effect associated with the spontaneous polarization. According to the definition of the piezoelectric coefficients d and e, the potential difference (in the converse effect) depends not only on the spontaneous polarization but also on the permittivity. Thus, a comparison of the spontaneous strain (or the spontaneous stress) with the field intensity does not yield a direct relationship between these quantities and the spontaneous polarization. If we describe the spontaneous piezoelectric effect in terms of the coefficients g and h, we find that the spontaneous strains or stresses are directly related to the charge on the electrodes of a capacitor, i.e., to the value of P or D.*

---

*The appearance of the spontaneous polarization P_s in a sample with short-circuited electrodes is equivalent to the acquisition of an induced charge given by the relationship $D = 4\pi P$. In view of this, the first and fifth equations in the system (IX.33), which may be used to describe the spontaneous piezoelectric effect, can be expressed in the form

$$t_j = -h_{ni}P_{sn},$$
$$r_i = g_{ni}P_{sn}. \qquad (IX.50)$$

For a sample with open-circuited electrodes or one which is well insulated and has a high resistivity, we find that $D = 0$, and the spontaneous polarization charge determines the electric field E:

$$E = -4\pi P_s/\varepsilon. \qquad (IX.50a)$$

In this case, the following relationships [Eq. (IX.33)] should be obeyed at the phase-transition point:

$$r_i = -4\pi d_{ni}P_{sn}/\varepsilon_n,$$
$$t_j = 4\pi e_{mj}P_{sm}/\varepsilon_m, \qquad (IX.50b)$$

and these relationships are found to be identical with the expressions in Eq. (IX.50) when Eq. (IX.37) is taken into account.

The piezoelectric properties of some ferroelectric crystals
have been investigated quite thoroughly although the applications
of these crystals are limited. We shall now consider the piezoelectric properties of the most typical ferroelectrics, representing
different classes of compounds.

A. Barium Titanate. The appearance of the spon-
taneous polarization in BaTiO$_3$ at 120°C is accompanied by a spontaneous electrostriction deformation, discussed in §2 of Chap. V
as well as in §5 of the present chapter. Since the paraelectric
modification has a center of inversion, barium titanate crystals
which are split into domains in the tetragonal (ferroelectric) modification do not exhibit piezoelectric properties (provided they
are not unipolar). The application of an alternating electric field
to such crystals produces a characteristic hysteresis in the dependence of the strain on the field, as shown in Fig. 250. The piezoelectric effect in a strongly polarized (single-domain) sample
is represented by lines 2-1 and 4-5 in the ideal curve of Fig. 250b.
Lines 4-6 and 2-3 also represent the piezoelectric effect in a single-domain sample, but they correspond to fields lower than the
coercive value (Fig. 250a). Lines 3-4 and 6-2 represent a reversal of the spontaneous polarization direction (Fig. 250a). In a
real crystal the hysteresis cycle becomes smoother and its shape
is of the type shown in Fig. 250c.

The application of an electric field along the spontaneous
polarization direction of multidomain samples of BaTiO$_3$ with 180°
domain walls produces an interesting phenomenon in the form of

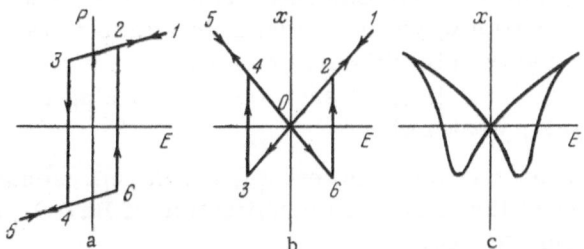

Fig. 250. Schematic representation of the converse piezoelectric effect in BaTiO$_3$: a) hysteresis loop of the dependence P(E) in the ideal case; b) ideal hysteresis loop of the
dependence of the strain on the field; c) butterfly-type
loop for a real crystal.

piezoelectric clamping of domains. Its origin is due to the fact that in weak fields the piezoelectric strain compresses domains with a given direction of polarization and expands domains with the opposite polarization. Such clamping reduces the permittivity measured in those states in which a crystal has zero or very weak polarization.

It is thus necessary to convert a ferroelectric crystal to the single-domain state before its piezoelectric properties can be investigated or used. A crystal can be converted completely or partially to the single-domain state by the application of a static electric field. It is worth mentioning that the domain structure of all the ferroelectric modifications of $BaTiO_3$ is centrosymmetric and that polarization is always required for the detection of the piezoelectric effect.

A centrosymmetric force, such as a mechanical stress or strain, cannot suppress the inversion center (center of symmetry) of a centrosymmetric domain configuration (this follows from the Curie principle of symmetry). This means that a $BaTiO_3$ crystal cannot be converted to the single-domain state simply by uniform strain. However, uniform mechanical forces can alter the domain structure (without suppressing the inversion center) of a crystal taken as a whole. It follows from symmetry considerations that when a crystal is deformed, for example, along the Y axis, the directions of polarization P_1 are oriented preferentially in a plane perpendicular to this axis. In this case, deformation does not produce the piezoelectric effect (i.e., polarization) but rotates the domains by 90°.

Bearing this in mind, our references to the piezoelectric effect in $BaTiO_3$ will always imply that a crystal is in the single-domain state and the piezoelectric effect is not associated with the domain switching, i.e., we shall consider the piezoelectric effect in relatively weak fields.

The piezoelectric coefficient g_{33}, which represents the piezoelectric effect of the tetragonal modification of $BaTiO_3$, can be estimated using Eq. (IX.73):

$$g_{ijm} = 2Q_{ijmn}P_n,$$

which can be written in the form

$$g_{33} = 2Q_{33}P_s.$$

TABLE 56. Values of the Principal Parameters of Tetragonal
BaTiO$_3$ (cgs esu) [16]

Tetragonal modification, t = 25°C

Single crystals				Permittivity ε of polarized ceramic
Elastic constants s, 10^{-12}	Piezoelectric coefficients d, 10^{-6}	Piezoelectric coefficients g, 10^{-8}	Permittivity ε	
$s_{11}^E = 0.805$	$d_{15} = 11.76$	$g_{15} = 5.07$	$\varepsilon_{11}^t = 2920$	$\varepsilon_{11}^t = 1436$
$s_{33}^E = 1.57$	$d_{31} = -1.04$	$g_{31} = 7.67$	$\varepsilon_{11}^r = 1970$	$\varepsilon_{11}^r = 1123$
$s_{12}^E = -0.235$	$d_{33} = 2.57$	$g_{33} = 19.17$	$\varepsilon_{33}^t = 168$	$\varepsilon_{33}^t = 1680$
$s_{13}^E = -0.524$			$\varepsilon_{33}^r = 110$	$\varepsilon_{33}^r = 1256$
$s_{44}^E = 1.84$				
$s_{66}^E = 0.884$				
$s_{11}^D = 0.725$				
$s_{33}^D = 1.08$				
$s_{12}^D = -0.315$				
$s_{13}^D = -0.326$				
$s_{44}^D = 1.24$				

Using the values given in Chap. V ($P_s = 57 \times 10^3$ cgs esu,
$Q_{33} = Q_{11} = 1.17 \times 10^{-12}$ cgs esu), we obtain

$$g_{33} = 13.3 \cdot 10^{-8} \text{ cgs esu,}$$

which is in good agreement with the experimental value which we
shall give later.

In accordance with the symmetry of the tetragonal modification of BaTiO$_3$ (group 4·m), its piezoelectric properties are described by the following set of piezoelectric strain coefficients
(Table 46):

$$\begin{pmatrix} 0 & 0 & 0 & 0 & d_{15} & 0 \\ 0 & 0 & 0 & d_{15} & 0 & 0 \\ d_{31} & d_{31} & d_{33} & 0 & 0 & 0 \end{pmatrix}.$$

The values of the piezoelectric, elastic, and dielectric parameters
of BaTiO$_3$ are listed in Table 56. It is evident from this table that
the piezoelectric strain coefficients d of barium titanate are at

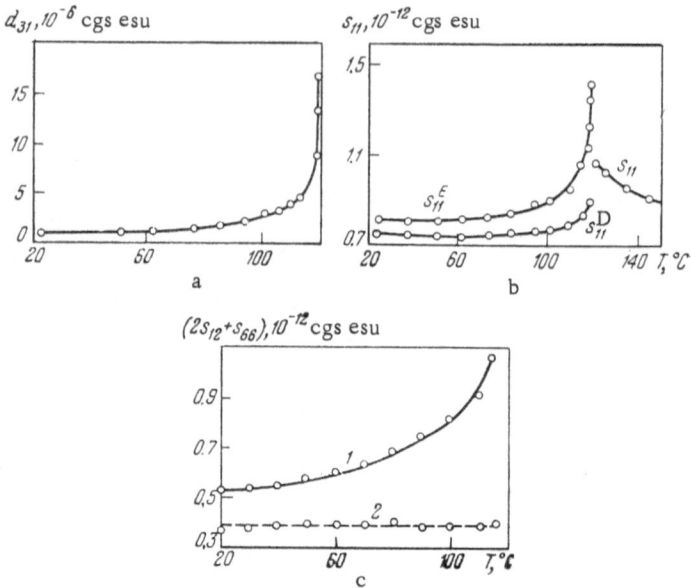

Fig. 251. Temperature dependences of the piezoelectric and elastic coefficients of BaTiO$_3$: a) piezoelectric strain coefficient d$_{31}$; b, c) elastic constants s measured in a constant electric field and for a constant polarization (curve 1 in Fig. 251c represents measurements in a constant E and curve 2 represents measurements in the case of a constant P) [18].

least an order of magnitude larger than the highest coefficients of linear piezoelectrics.

The large difference between the values of s^E and s^D as well as between the values of ε^t and ε^r indicates that the electromechanical coupling coefficient k of BaTiO$_3$ crystals is large (it can reach 67% for some vibration modes).

There are no published data on the temperature dependences of the coefficients g$_{ij}$, but it has been reported that the coefficient d$_{31}$ increases near the transition temperature for reasons given earlier in this subsection; the temperature dependence of d$_{31}$ is given in Fig. 251a.

Figures 251b and 251c show the temperature dependences of the elastic constants s$_{ij}$ of BaTiO$_3$ crystals. The rapid rise of

s_{11}^E near the Curie point can be explained in the same way as the rise of the piezoelectric coefficients d and e, i.e., by the reaction of a crystal to a change in its permittivity. We have mentioned earlier that $s^D < s^E$. The rise of the permittivity near the Curie point increases the difference between s^D and s^E of a short-circuited sample (or a sample subjected to a constant voltage which should remain the same at all temperatures). This is because a crystal must be supplied with an increasing charge as it approaches the Curie temperature. The state of a crystal seems to depart even further from its isolated state (D = 0) and the value of s^E increases. The weak rise of s_{11}^D is evidently associated with the multidomain state of this crystal near the Curie point.

Comprehensive and reliable information on the piezoelectric properties of other oxygen-octahedral ferroelectric crystals is lacking. Practically the only information available on these crystals is that $PbTiO_3$ has piezoelectric coefficients which are even higher than those of barium titanate (this is evidently related to the higher value of the spontaneous polarization of lead titanate). This means that it would be desirable to grow single crystals of $BaTiO_3 - PbTiO_3$ solid solutions containing up to 5-8% of the lead compound.

We shall not consider the piezoelectric effect in oxygen-octahedral antiferroelectrics and weak ferroelectrics, confining ourselves to the statement that, in principle, this effect is possible in such substances. There are some erroneous suggestions that the piezoelectric effect is impossible in antiferroelectrics, and the absence of the piezoelectric effect beyond the phase transition point has even been used as evidence that the transition is antiferroelectric. The piezoelectric effect can occur in antiferroelectrics provided they have no inversion center (center of symmetry) before the transition. Such crystals do not have an inversion center after the transition and the formation of a domain structure. The piezoelectric effect should occur also in some antiferroelectrics which have an inversion center in the paraelectric modification. This is because an ideal domain structure is difficult to establish in such crystals and, therefore, they may consist of domain arrays without an inversion center. The rules governing the piezoelectric effect in weak ferroelectrics are the same as those for ordinary ferroelectrics.

The piezoelectric properties of solid solutions of oxygen-octahedral ferroelectrics, in which the main component is usually $BaTiO_3$, have found a wide range of practical applications. Admixtures in these solid solutions are $PbTiO_3$, $CaTiO_3$, $SrTiO_3$, $PbZrO_3$, and many other compounds. These solutions are used in the ceramic form. The piezoelectric and other properties of ceramics do not differ greatly from the properties of single crystals, and the numerical values are similar (Table 56).

Piezoelectric ceramics made of oxygen-octahedral ferroelectrics have several valuable properties: high strength, insolubility in water, and stability. Piezoelectric ceramics can be machined quite easily and can be used to make large elements of various shapes. Strong piezoelectric and nonlinear properties of oxygen-octahedral ferroelectrics are so important that special research and industrial organizations, concerned solely with these materials, have been established in many countries. Descriptions of a large number of piezoelectric ceramics and their properties are given in Martin's book [19].

It is worth mentioning that polycrystalline piezoelectric ferroelectrics, synthesized by ceramic techniques, become piezoelectric textures after polarization in an electric field (§1 in the present chapter).

B. Potassium Dihydrogen Phosphate. The paraelectric modification of potassium dihydrogen phosphate (which exists above −151°C) belongs to the tetragonal scalenohedral symmetry class $\bar{4} \cdot m$ and exhibits piezoelectric properties. The matrix of the piezoelectric strain coefficients of this class, derived using Table 46 and the coordinate system shown in Fig. 252, is of the form

$$\begin{pmatrix} 0 & 0 & 0 & d_{14} & 0 & 0 \\ 0 & 0 & 0 & 0 & d_{14} & 0 \\ 0 & 0 & 0 & 0 & 0 & d_{36} \end{pmatrix}. \qquad \text{(IX.51)}$$

The experimental value of the piezoelectric coefficient g_{36} of potassium dihydrogen phosphate is 5.0×10^{-7} cgs esu.

The appearance of the spontaneous polarization at −151°C is accompanied by a spontaneous shear r_6, which is given by the relationship [Eq. (V.33)]:

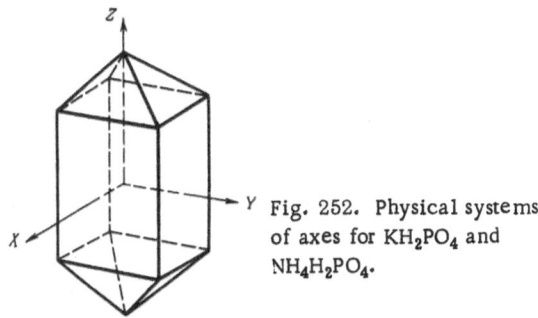

Fig. 252. Physical systems of axes for KH_2PO_4 and $NH_4H_2PO_4$.

$$r_6 = g_{36}P_3. \qquad (IX.52)$$

Comparison of the values of r_6 and P_3, given in Chapter V, yields $g_{36} = 5.5 \times 10^{-7}$ cgs esu, which is in good agreement with the experimental value for the paraelectric modification. It must be stressed that throughout the ferroelectric range of temperatures (including the vicinity of the Curie point) P_s and r_6 remain proportional. This can be seen in Fig. 128. Since g_{36} does not exhibit anomalies on passing through the Curie point it follows that this coefficient is the true coefficient of a crystal which is not associated with anomalies of its dielectric properties. This cannot be said about the piezoelectric strain coefficient d_{36}, which increases on approach to the Curie point and obeys the Curie–Weiss law, which is of the form

$$d_{36} = d_{36}^0 + \frac{B}{T - T_c}, \qquad (IX.53)$$

where $B = 1.26 \times 10^{-4}$ cgs esu and $d_{36}^0 = -8 \times 10^{-8}$ cgs esu. The piezoelectric strain coefficient d_{14} also shows anomalous behavior (but not so strongly pronounced) near the Curie point, associated with a weak anomaly of ε_a (Fig. 179). The room–temperature value of the coefficient d_{14} is 4×10^{-8} cgs esu.

The appearance of the spontaneous polarization along the c axis reduces the symmetry of potassium dihydrogen phosphate: $\bar{4} \cdot m \rightarrow 2 \cdot m$. This orthorhombic symmetry is exhibited by a single-domain crystal or by its individual domains. The piezoelectric properties of the orthorhombic modification (class $2 \cdot m$) are described by the following matrix of the piezoelectric strain co-

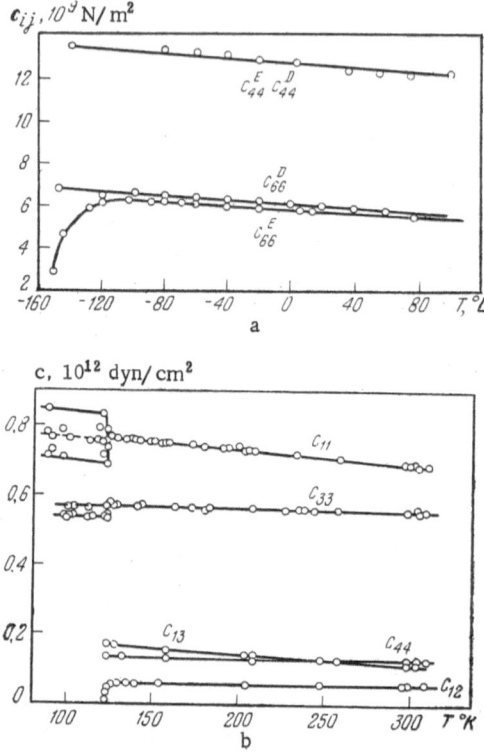

Fig. 253. Temperature dependences of the elastic moduli c_{ij} of potassium dihydrogen phosphate.

efficients (Table 46):

$$\begin{pmatrix} 0 & 0 & 0 & 0 & d_{15} & 0 \\ 0 & 0 & 0 & d_{24} & 0 & 0 \\ d_{31} & d_{31} & d_{33} & 0 & 0 & 0 \end{pmatrix}. \qquad (IX.54)$$

The value of the piezoelectric coefficient g_{33} in the ferroelectric region, estimated from the electrostrictive strain ($g_{33} = 8 \times 10^{-8}$ cgs esu) is given in Chap. V. This value of g_{33} and the temperature dependence of the permittivity can be used to calculate the piezoelectric strain coefficient d_{33} and its temperature dependence. The values of the coefficients d_{31} and d_{32} are close to the value of d_{36} (although they differ somewhat). This can be understood quite easily by taking into account the fact that the X and Y axes of the single-domain state are rotated by 45° relative to the X and Y axes of the paraelectric modification. Hence, we obtain the following approximate relations:

$$r_{6\,pyro} \cong 2r_{1\,ferro} \cong 2r_{2\,ferro}$$

The spontaneous polarization is related to the spontaneous shear r_6. This means that the elastic coefficients c and s, related to r_6 and measured in a constant field (E = 0), should exhibit anomalies near the Curie point. According to the generalized Hooke's law (Table 50), these coefficients are c_{66}^E and s_{66}^E. The temperature dependences of the elastic moduli c_{66}^E, c_{66}^D, c_{44}^D, and c_{44}^E are given in Fig. 253a. It is found that the difference $(s_{66}^E - s_{66}^D)$ obeys the Curie−Weiss law. The values and the temperature dependences of other elastic moduli are given in Fig. 253b.

The principal elastic, piezoelectric, and dielectric parameters of potassium dihydrogen phosphate are listed in Table 57.

When we attribute anomalies of the coefficients d^E and c^E (or s^E) to the anomalies of the permittivity of potassium dihydrogen phosphate, we must specify the relevant permittivity (ε^t or ε^r). It can be shown that the anomalies of the piezoelectric and other coefficients are associated with the anomalies of the permittivity of a clamped crystal ε^r. Calculations and experiments show that the Curie−Weiss temperature of a clamped crystal is 4°C lower than the corresponding temperature of a free crystal. Hence, we may expect the Curie temperature of a clamped crystal to be 4°C lower than the Curie temperature of a free crystal. However, experiments show that the Curie temperatures are the same in the clamped and free states. This shows that the differ-

TABLE 57. Values of the Principal Parameters of Potassium Dihydrogen Phosphate (cgs esu)

Elastic constants, s, 10^{-12} (25°C)	Elastic moduli c, 10^{10} (25°C)	Piezoelectric coefficients d and e (20°C)	Permittivity ε (20°C)	Thermal expansion coefficients α, 10^{-6} deg^{-1}
$s_{11} = 1.65$	$c_{11} = 86.8$	$d_{11} = 4.2 \cdot 10^{-8}$	$\varepsilon_{11} = 4.6$	$\alpha_1 = \alpha_2 = 36.1$
$s_{12} = -0.4$	$c_{12} = 40.2$	$d_{36} = 69.6 \cdot 10^{-8}$	$\varepsilon_{33} = 21.8$	$\alpha_3 = 41.1$
$s_{13} = -0.75$	$c_{13} = 47.6$	$e_{14} = 0.53 \cdot 10^1$		(from −40
$s_{23} = 2.0$	$c_{23} = 85.7$	$e_{36} = 4.26 \cdot 10^1$		to + 80°C)
$s_{44} = 7.9$	$c_{44} = 12.7$			
$s_{66} = 16.6$	$c_{66} = 6.0$			

ence between the permittivities of clamped and free crystals is independent of the temperature. Using the equations for the piezo-electric effect (IX.24), (IX.25), (IX.27)–(IX.32), we obtain an expression for the difference $\beta_3^r - \beta_3^t$ [Eq. (IX.40)]:

$$\beta_3^r - \beta_3^t = \frac{(d_{36})^2}{(\varepsilon_3^t)^2 \, s_{66}^D} . \qquad\qquad \text{(IX.55)}$$

At the Curie point of a free crystal ($\beta^t \to 0$) the reciprocal of the permittivity of a clamped crystal (ε^r) is found from the expression

$$\beta_3^r = \frac{(d_{36})^2}{(\varepsilon_3^t)^2 \, s_{66}^D} \qquad\qquad \text{(IX.56)}$$

and amounts to about 7×10^{-5}. In the last expression the ratio $(d_{36}/\varepsilon_3^t)^2$ determines, on the basis of Eq. (IX.37), the piezoelectric coefficient g which is independent of temperature. Thus, it follows from Eq. (IX.56) that the anomalies of the elastic constant s_{66}^D are due to anomalies in the behavior of the permittivity of a clamped crystal ($\varepsilon_{66}^r \propto s_{66}^D$).

C. Triglycine Sulfate. At room temperature triglycine sulfate is monoclinic (prismatic class 2:m). The presence of an inversion center in this crystal indicates that the paraelectric modification should not be piezoelectric. The appearance of the spontaneous polarization in triglycine sulfate at 49°C is accompanied by an electrostrictive deformation. The compound remains monoclinic but its symmetry changes to the dihedral axial class 2, whose piezoelectric properties are described by the fol-

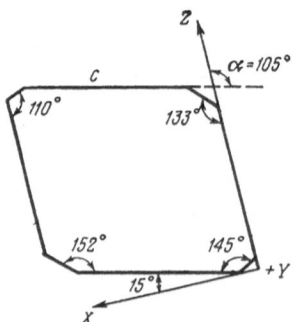

Fig. 254. Physical system of coordinates for triglycine sulfate.

TABLE 58. Values of the Principal Parameters
of Triglycine Sulfate (cgs esu) [10]

Elastic constants s, 10^{-12}	Elastic moduli c, 10^{10}	Piezoelectric coefficients d, 10^{-8} (23°C)	Permittivity ε
$s_{11} = 4.22$	$c_{11} = 4.28$	$d_{21} = -44.2$	$\varepsilon_{11} = 8.7$
$s_{22} = 6.49$	$c_{22} = 33.2$	$d_{22} = -83,3$	$\varepsilon_{22} = 25$
$s_{33} = 9.75$	$c_{33} = 28.5$	$d_{23} = 136.7$	$\varepsilon_{33} = 4.7$
$s_{41} = 10.51$	$c_{44} = 9.67$	$d_{25} = 45.3$	$\varepsilon_{13} = 0.5$
$s_{55} = 10.37$	$c_{55} = 11.8$	$d_{14} = 13.4$	
$s_{66} = 16.00$	$c_{66} = 6.35$	$d_{16} = -14.7$	
$s_{23} = -4.61$	$c_{23} = 20.6$	$d_{34} = 0.5$	
$s_{13} = -2.90$	$c_{13} = 21,6$	$d_{36} = 11.6$	
$s_{12} = -0.43$	$c_{12} = 17.6$		
$s_{15} = 0.13$	$c_{15} - 18.5$		
$s_{25} = -2.03$	$c_{25} = -1.16$		
$s_{46} = -2,81$	$c_{35} = 3.4$		
	$c_{46} = -1,0$		

lowing matrix which consists of eight independent piezoelectric
strain coefficients (the system of coordinates is shown in Fig.
254):

$$\begin{pmatrix} 0 & 0 & 0 & d_{14} & 0 & d_{16} \\ d_{21} & d_{22} & d_{23} & 0 & d_{25} & 0 \\ 0 & 0 & 0 & d_{34} & 0 & d_{36} \end{pmatrix}. \qquad \text{(IX.57)}$$

Comparison of the spontaneous deformation and polarization
of a crystal makes it possible to estimate the coefficients g_{ik}
(Chap. V):*

$$g_{22} = -25.2 \cdot 10^{-8},$$
$$g_{21} = -59.2 \cdot 10^{-8},$$
$$g_{23} = 15.2 \cdot 10^{-8} \text{ cgs esu.}$$

In accordance with the principle which states that, after splitting into domains, a crystal reverts to the symmetry of its paraelectric modification, a multidomain crystal of triglycine sulfate
(unless unipolar) cannot exhibit piezoelectric properties. However, experiments show that triglycine sulfate crystals (more accurately, samples prepared from such crystals) are usually unipolar and exhibit piezoelectric properties. Naturally, the intrinsic
properties of a crystal are obtained only in the single-domain state.

---

*The values of the coefficient g_{22} are estimated for room temperature and those of
g_{21} and g_{23} are estimated at 37°C.

Many papers have been published on the piezoelectric and elastic properties of triglycine sulfate. The fullest information is given by Sil'vestrova [10] and we shall base our discussion on her data. Table 58 gives the room-temperature values of the principal parameters of triglycine sulfate.

The temperature dependences of the elastic constants s_{ij} show that, since they are measured at a constant field intensity, they depend strongly on the temperature near the Curie point (a peak of s^E lies a little below T_c and there is a rapid variation at T_c). On the other hand, the elastic constants measured at a constant value of the electric induction show practically no anomalies in the phase-transition region. This applies to the constants s_{11}^D, s_{22}^D, and s_{33}^D

The electromechanical coupling coefficient of triglycine sulfate depends on the degree of polarization of a sample and on its temperature. Thus, this coefficient for a Y-cut slab (a ferroelectric cut), oriented along the Z axis, increases from 25 to 54% on approach to T_c from the room-temperature side and it falls strongly at the Curie point. When polarizing fields $E \approx 200\text{-}300$ V/cm are applied to a slab of the same cut and the same orientation, the room-temperature value of k reaches about 30-35%. A change in the polarizing field intensity switches the polarization of the sample and k passes through zero, but in a full cycle it traces out a butterfly-shaped curve (Fig. 250c).

The application of a static field to the paraelectric modification of triglycine sulfate has little effect on its elastic constants (the values of s increase by about 3% in a field of 1000 V/cm). In the ferroelectric range of temperatures the application of fields producing domain reorientation raises considerably the values of the elastic constants s, which increase with increasing field intensity. This increase amounts to about 15% in a field of 1000 V/cm.

Investigations of the temperature dependences of the piezoelectric properties show that the piezoelectric strain coefficients increase on approach to the Curie point, and this increase is approximately proportional to the square of the temperature; at the Curie point these coefficients decrease strongly. Similar anomalies are exhibited also by the temperature dependences of other piezoelectric coefficients.

The temperature dependence of the piezoelectric strain coefficient d_{22} is shown in Fig. 255. It is evident from this figure

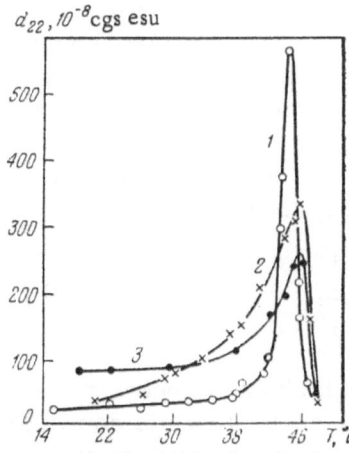

$d_{22}, 10^{-8}$ cgs esu

Fig. 255. Temperature dependence of the piezoelectric strain coefficient d_{22} of triglycine sulfate, determined using different intensities of a polarizing field $E_=$ (V/cm): 1) 0; 2) 300; 3) 600. $E_\sim = 50$ V/cm [13].

that $d_{22} = 80 \times 10^{-8}$ cgs esu for the polarized state at room temperature, which is in good agreement with the value listed in Table 58, found by a different method.

It must be mentioned that general considerations indicate that the anomalies of the piezoelectric strain coefficients d of triglycine sulfate should (in principle) differ from the anomalies of d in crystals which are not centrosymmetric in the paraelectric modification (potassium dihydrogen phosphate and Rochelle salt). This difference between the anomalies follows from the fact that, in the case of triglycine sulfate (and similar crystals), all the piezoelectric coefficients vanish on transition to the paraelectric modification, whereas in the case of potassium dihydrogen phosphate some piezoelectric coefficients do not vanish even in the paraelectric phase.

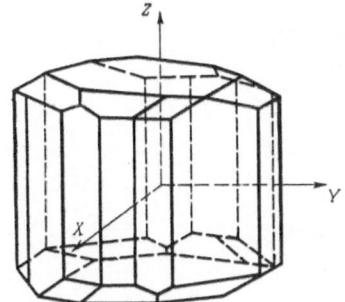

Fig. 256. Rochelle salt crystal and its physical system of coordinates.

D. Rochelle Salt. The paraelectric modification of
Rochelle salt (which exists above 24°C and below −18°C) belongs
to the rhombic tetrahedral class 2:2 of the orthorhombic system.
Since this modification is noncentrosymmetric, Rochelle salt exhibits piezoelectric properties and the matrix of its piezoelectric
strain coefficients has three independent components (Table 46
and the coordinate system shown in Fig. 256):

$$\begin{pmatrix} 0 & 0 & 0 & d_{14} & 0 & 0 \\ 0 & 0 & 0 & 0 & d_{25} & 0 \\ 0 & 0 & 0 & 0 & 0 & d_{36} \end{pmatrix}. \tag{IX.58}$$

The spontaneous polarization, which exists between −18°C
and 24°C, makes this compound (more accurately, its domains)
monoclinic of the symmetry class 2. Using the system of coordinates shown in Fig. 256, the matrix of the piezoelectric strain coefficients of a single-domain crystal of Rochelle salt can be written
in the form

$$\begin{pmatrix} d_{11} & d_{12} & d_{13} & d_{14} & 0 & 0 \\ 0 & 0 & 0 & 0 & d_{25} & d_{26} \\ 0 & 0 & 0 & 0 & d_{35} & d_{36} \end{pmatrix}. \tag{IX.59}$$

The matrices of the other piezoelectric coefficients are similar.

The new piezoelectric coefficients due to the spontaneous
polarization can be calculated using the corresponding electrostrictive coefficients employing Eq. (IX.73):

$$g_{ijm} = 2Q_{ijmn}P_n.$$

Using this equation and the values of the electrostrictive coefficients, we can estimate (Chap. V) the coefficients g_{ij} (in cgs esu)
near the upper Curie point

$$g_{11} = -9.44 \cdot 10^{-8},$$
$$g_{12} = 2.00 \cdot 10^{-8},$$
$$g_{13} = -2.78 \cdot 10^{-8};$$

near the lower Curie point

$$g_{11} = 10.7 \cdot 10^{-8},$$
$$g_{12} = -2.12 \cdot 10^{-8},$$
$$g_{13} = 3.0 \cdot 10^{-8};$$

and in the middle of the ferroelectric range of temperatures

$$g_{11} = -6.6 \cdot 10^{-8},$$
$$g_{12} = 1.3 \cdot 10^{-8},$$
$$g_{13} = -1.8 \cdot 10^{-8}.$$

Using the permittivities ε^t and the electrostrictive coefficients, we can estimate the piezoelectric strain coefficients d_{11}, d_{12}, and d_{13} (in the middle of the ferroelectric range of temperatures):

$$d_{11} \cong -1 \cdot 10^{-6},$$
$$d_{12} \cong 2 \cdot 10^{-7},$$
$$d_{13} \cong -3 \cdot 10^{-7} \text{ cgs esu.}$$

The experimentally determined temperature dependence of the coefficient d_{11} is shown in Fig. 257a for various values of the polarizing fields. Anomalies in the behavior of the piezoelectric strain coefficients near the Curie point and the influence of the electric field can be explained as for other compounds, i.e., by assuming that the piezoelectric strain coefficients are related to the spontaneous polarization or, more precisely, they are governed entirely by this polarization. In the middle of the ferroelectric range of temperatures the room–temperature value of d_{11} for a

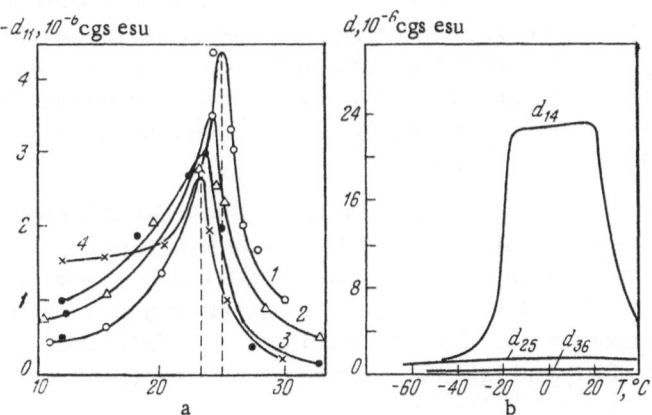

Fig. 257. Temperature dependences of the piezoelectric strain coefficients of Rochelle salt. a) Coefficient d_{11} of the monoclinic modification, measured using various values of the polarizing field E (V/cm): 1) 0; 2) 69; 3) 91; 4) 304 [11]. b) Coefficients d_{14}, d_{25}, and d_{36} [21].

polarized sample is estimated to be $(1-1.5) \times 10^{-6}$ cgs esu. Similar measurements give $d_{12} = 3.8 \times 10^{-7}$ cgs esu and $d_{13} = -3.78 \times 10^{-7}$ cgs esu, which are in satisfactory agreement with the values of the coefficients d_{11}, d_{12}, and d_{13} calculated from the electrostrictive coefficients (given in the preceding paragraph).

The piezoelectric strain coefficients d_{14}, d_{25}, and d_{36} (particularly the first) have been investigated most thoroughly using a variety of methods. The experimentally determined temperature dependences of these three moduli are given in Fig. 257b. It is evident from this figure that the coefficient d_{14} exhibits a strong anomaly in the ferroelectric range of temperatures. The absence of a drop in d_{14} in the middle of the ferroelectric range (expected by analogy with the temperature dependence of ε_1) is attributed by Valasek [21] to the domain switching in strong fields (this point is discussed later). Above the upper Curie point the coefficient d_{14} depends on temperature in accordance with the Curie–Weiss law.

It has been demonstrated experimentally that the piezoelectric coefficients g_{14} and h_{14} of Rochelle salt are practically independent of the temperature (including the regions of both Curie points) and of the field intensity, whereas the coefficients d_{14} and e_{14} depend on the temperature and field intensity. This shows that the coefficients g_{14} and h_{14} represent the "intrinsic" properties. The absence of anomalies in the behavior of the coefficient g_{14} follows directly from the similarity of the temperature dependences $P_s(T)$ (Fig. 118) and of the extinction angle of domains, which is proportional to the strain r_4 (Fig. 97). This comparison (Chap. V) gives $g_{14} = 4.0 \times 10^{-7}$ cgs esu for a wide range of temperatures, including both Curie points.

The elastic moduli c_{44}* and c_{66} of Rochelle salt measured at constant electric induction (polarization) vary only slightly in the temperature range which includes both Curie points (Fig. 258). On the other hand, the elastic moduli, measured in a constant field, increase strongly near the Curie points. The difference between the values of s_{44}^E and s_{44}^D also obeys the Curie–Weiss law.

---

*The elastic modulus c_{44} is related to the spontaneous shear of Rochelle salt (see the matrix of the coefficients in the generalized Hooke's law in Table 50). All the elastic moduli c_{ij} of Rochelle salt, which are not associated with the polarization along the ferroelectric axis, show no anomalies of any kind at the Curie point and depend weakly on temperature.

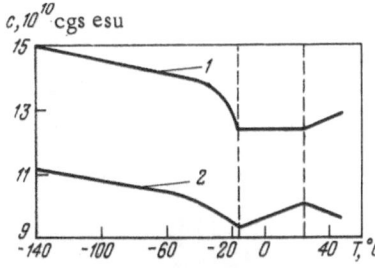

Fig. 258. Temperature dependences of the elastic moduli c_{44}^D (1) and c_{66}^D (2) of Rochelle salt (measured using a holder with an air gap).

Table 59 lists the principal electrical, piezoelectric, and dielectric parameters of Rochelle salt. The values given in that table refer to room temperature (which is not specified more precisely). It would be desirable to know the temperature more exactly because the upper Curie point is close to room temperature and some of the parameters depend strongly on the temperature near this point. It is evident from Table 59 that Rochelle salt has very strong piezoelectric properties, particularly when they are represented by the coefficients d and g. The large difference between the values of ε^t and ε^r is responsible for the high values of the electromechanical coupling coefficient k, exceeding 50% for some vibration modes.

Rochelle salt crystals which are split into domains still exhibit piezoelectric properties. This makes it possible to switch the polarization of these crystals by the application of a mecha-

TABLE 59. Values of the Principal Parameters of Rochelle Salt (cgs esu)

Elastic constants s, 10^{-12}	Elastic moduli c, 10^{10}	Piezoelectric coefficients d and g, 10^{-8}	Piezoelectric coefficients e and h, 10^4	Permittivity ε	Thermal expansion coefficients α, 10^{-6} deg^{-1}
$s_{11} = 5.2$	$c_{11} = 40$	$d_{14} = 1150$	$e_{14} = 140$	$\varepsilon_{11}^t = 480$	$\alpha_1 = 58.3$
$s_{12} = -2.1$	$c_{12} = -30$	$d_{25} = -160$	$e_{25} = -5$	$\varepsilon_{22}^t = 12$	$\alpha_2 = 35.5$
$s_{13} = -2.0$	$c_{13} = -40$	$d_{36} = 35$	$e_{26} = 3.5$	$\varepsilon_{33}^t = 10$	$\alpha_3 = -136.1$
$s_{22} = 3.4$	$c_{22} = 50$	$g_{14} = 31$	$h_{14} = 7.58$	$\varepsilon_{11}^r = 220$	
$s_{23} = -1.3$	$c_{23} = -30$	$g_{25} = -170$	$h_{25} = -5.8$	$\varepsilon_{22}^r = 11$	
$s_{33} = 3.2$	$c_{33} = 60$	$g_{36} = 44$	$h_{36} = 4.8$	$\varepsilon_{33}^r = 9.8$	
$s_{44} = 20$	$c_{44} = 10$				
$s_{55} = 32$	$c_{55} = 3$				
$s_{66} = 10$	$c_{66} = 10$				

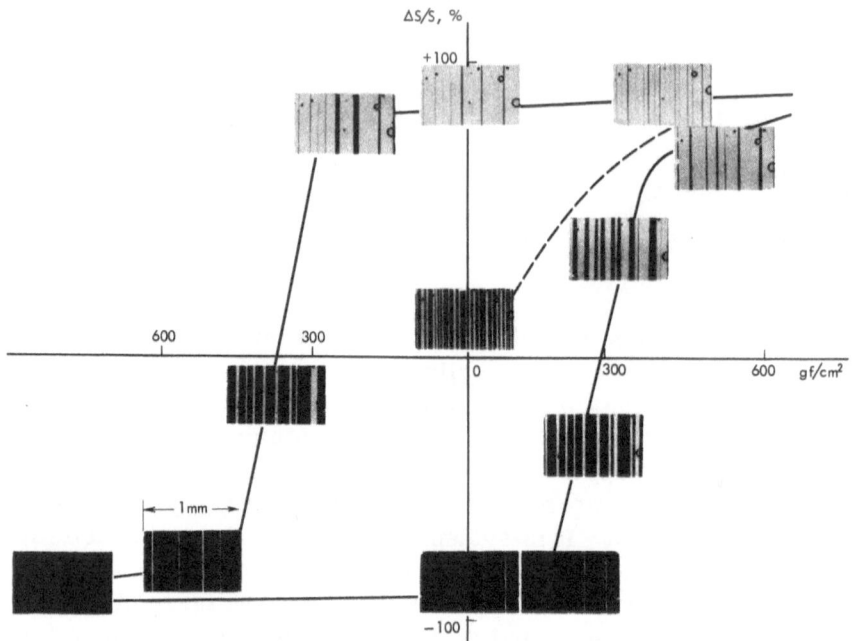

Fig. 259. Cyclic polarization reversal of domains in an X-cut sample of Rochelle salt under the action of a mechanical load [3]. The initial state is shown at the center of the figure.

nical stress t_4 (or a strain r_4). Changes in the domain structure under the action of alternating mechanical stresses produce a characteristic dependence of the polarization (represented by the ratio of the areas of components of domains of opposite orientation) on the mechanical stress, which is in the form of a hysteresis loop (Fig. 259). Comparison of the values of the mechanical stresses and electric fields which produce the same domain configuration (for $P = 0$, $E = E_c$, where E_c is the coercive field) makes it possible to estimate the value of the piezoelectric coefficient e_{14}, which is found to be in good agreement with the value deduced from electrical measurements (ignoring changes in the domain structure). Hence, we may conclude that the piezoelectric properties of single-domain and multidomain Rochelle salt crystals are practically identical.

E . SbSI.   Ferroelectric properties have recently been discovered in a large number of photoconducting compounds of elements of groups V, VI, and VII (V − Sb; VI − S, Se; VII − Cl, Br, I). These are the first compounds known to exhibit both ferroelectric and photoconducting properties. This point is very important because it gives additional information on the nature of the spontaneous polarization. Moreover, these compounds exhibit strong piezoelectric properties.

The most thoroughly investigated of these compounds is SbSI. Above 22°C this compound is paraelectric and orthorhombic (rhombic bipyramidal class m·2:m). The spontaneous polarization appears along the c axis so that after transition to the ferroelectric state these crystals (more exactly, the domains) assume the symmetry of class 2·m.

Table 60 lists the principal elastic, piezoelectric, and dielectric parameters of this compound at 17.9°C, i.e., at a temperature close to that at which the coefficient d_{33} reaches its maximum value.

F. Ammonium Dihydrogen Phosphate. Ammonium dihydrogen phosphate, $NH_4H_2PO_4$, is the only antiferroelectric (or the group of substances exhibiting a phase transition involving orderable structure elements) whose piezoelectric properties have been investigated. Unfortunately, these properties have been determined only for the paraelectric modification. In this modification (above −125°C) ammonium dihydrogen phosphate belongs, like potassium dihydrogen phosphate, to the tetragonal scalenohedral class $\bar{4}$·m of the tetragonal system and exhibits piezoelectric properties. The symmetry of domains along one of the antipolarization axes of this compound is 2:2 (this domain symmetry does not have an inversion center), but the symmetry of the whole crystal (split into domains) is, of course, $\bar{4}$·m. At the phase transition

TABLE 60. Values of the Principal Parameters
of SbSI (cgs esu)

Elastic moduli c, 10^{10}	Piezoelectric coefficients d, e, and coefficient k	Permittivity ε
$c_{33}^E = 15.3$	$d_{39} = 54.5 \cdot 10^{-6}$	$\varepsilon_{33}^t = 7750$
$c_{33}^D = 57.9$	$e_{33} = 830 \cdot 10^4$	$\varepsilon_{33}^r = 2046$
	$k \cong 86\%$	

TABLE 61. Values of the Principal Parameters of Ammonium
Dihydrogen Phosphate at 20°C (cgs esu)

Elastic constants s, 10^{-12}	Elastic moduli c, 10^{10}	Piezoelectric coefficients d and e	Permittivity ε	Thermal expansion coefficients α, 10^{-6} $\deg^{-1}$
$s_{11} = 1.81$	$c_{11} = 67.6$	$d_{14} = 5.27 \cdot 10^{-8}$	$\varepsilon_{11} = 57.6$	$\alpha_1 = \alpha_2 = 30.1$
$s_{12} = 0.19$	$c_{12} = 5.9$	$d_{36} = -145.0 \cdot 10^{-8}$	$\varepsilon_{33} = 14.0$	$\alpha_3 = 4.1$
$s_{13} = -1.18$	$c_{13} = 20.0$	$e_{14} = 0.46 \cdot 10^4$		
$s_{33} = 4.35$	$c_{33} = 33.8$	$e_{36} = -3.81 \cdot 10^4$		
$s_{44} = 11.53$	$c_{44} = 8.7$			
$s_{66} = 16.46$	$c_{66} = 6.1$			

point ($-125°C$) ammonium dihydrogen phosphate crystals are subjected to such strong stresses that they disintegrate into a powder. This process is irreversible and, therefore, it is impossible to investigate the piezoelectric properties of ammonium dihydrogen phosphate crystals at the transition temperature and below it.

Table 61 lists the principal elastic, piezoelectric, dielectric, and thermal parameters of ammonium dihydrogen phosphate. The temperature dependences of the elastic constants are given in Fig. 260. The electromechanical coupling coefficient of Z-cut (45°) samples is 30%.

The reported results indicate relatively strong piezoelectric properties of ammonium dihydrogen phosphate and they also show that these properties are fairly stable. This is why ammonium dihydrogen phosphate is used more widely in piezoelectric applications than other water-soluble crystals.

More detailed descriptions of the piezoelectric properties of linear piezoelectrics and of ferroelectrics discussed in the present book as well as of other crystals are given in the books of Cady [7], Mason [8], and Petrzilka et al. [20]. These books describe also various applications of the piezoelectric effect.

§5. Electrostriction

A. General Information. Electrostriction is a phenomenon which gives rise to a strain r_{ij}, which is proportional to the square of an applied electric field (E^2):

$$r_{ij} = R_{ijmn}E_m E_n. \qquad \text{(IX.60a)}$$

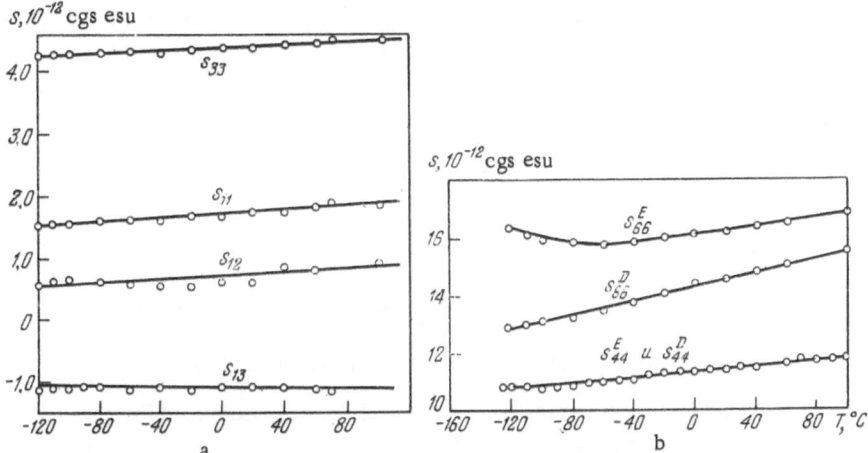

Fig. 260. Temperature dependences of the elastic constants of an $NH_4H_2PO_4$ crystal: a) elastic constants s_{11}, s_{12}, s_{13}, and s_{33}; b) elastic constants s_{66} and s_{44}.

Because of the quadratic dependence, the strain does not change its sign when the direction of the field is reversed and, therefore, in alternating electric fields of frequency f the frequency of elec-trostrictive deformation of a dielectric is $2f$. Electrostriction is due to the action of an electric field on charges of atoms and mole-cules in a dielectric, and it is observed in all dielectrics (solid, liquid, and gaseous) irrespective of their symmetry. Considera-tion of the possible relationships between the strain $[r_{ij}]$ and stress $[t_{ij}]$ tensors and the electric field E and polarization P vec-tors yields three additional equations

$$
\begin{aligned}
r_{ij} &= Q_{ijmn}P_mP_n, \\
t_{ij} &= G_{ijmn}P_mP_n, \\
t_{ij} &= H_{ijmn}E_mE_n.
\end{aligned}
\qquad \text{(IX.60b)}
$$

The coefficients R, Q, G, and H in Eqs. (IX.60a) and (IX.60b) are known as electrostrictive coefficients and are fourth-rank tensors. The meaning of these coefficients is evident from the equations themselves. Using a single index for the components of the tensors $[r_{ij}]$ and $[t_{ij}]$, as well as a notation of the type $P_1P_1 = P_1^2$, $P_2P_2 = P_2^2$, $P_3P_3 = P_3^2$, $P_2P_3 = P_3P_2 = P_4^2$, $P_3P_1 = P_1P_3 = P_5^2$, $P_1P_2 = P_2P_1 = P_6^2$, we obtain a double-index form of the tensors R, Q, G, and H

TABLE 62. Forms of the Matrices of Electrostrictive Coefficients
H, G, R, and Q of Crystals

Notation used for all matrices

·   zero coefficient

•   nonzero coefficient

•—• equal coefficients

•—o coefficients numerically equal but opposite in sign

Notation used for coefficients R and Q

⊙   coefficient equal to twice the heavy–dot coefficient
to which it is joined

◎   coefficient opposite in sign and numerically equal
to twice the heavy–dot coefficient to which it is
joined

✕ $2(R_{11} - R_{12})$; $2(Q_{11} - Q_{12})$

Notation used for coefficients G and H

⊙   coefficient equal to twice the heavy–dot coefficient
to which it is joined

◎   coefficient opposite in sign and numerically equal
to twice the heavy–dot coefficient to which it is
joined

✕ $(G_{11} - G_{12})$; $(H_{11} - H_{12})$

Triclinic System
Both classes

(36)

TABLE 62 (Continued)
Monoclinic System
All classes

2 ∥ Y; ⊥ m ∥ Y

2 ∥ Z; ⊥ m ∥ Z

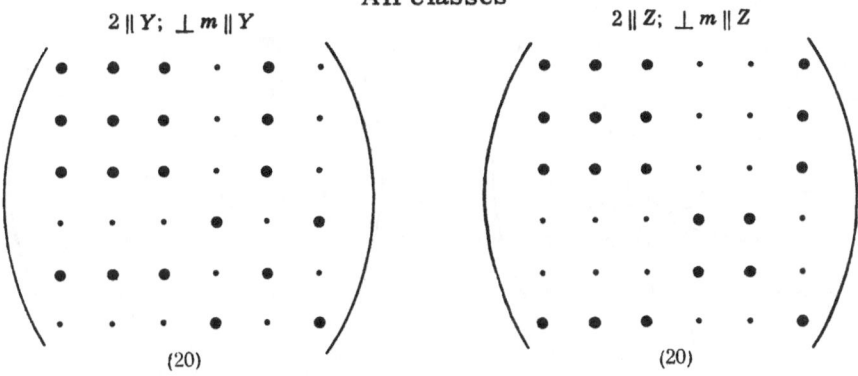

(20)

(20)

Orthorhombic System
All classes

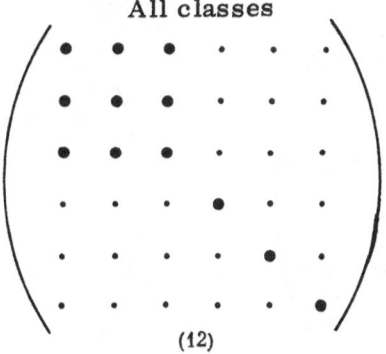

(12)

Tetragonal System

Classes 4, 4̄, 4 : m

Classes 4·m, 4̄·m, 4 : 2, m·4 : m

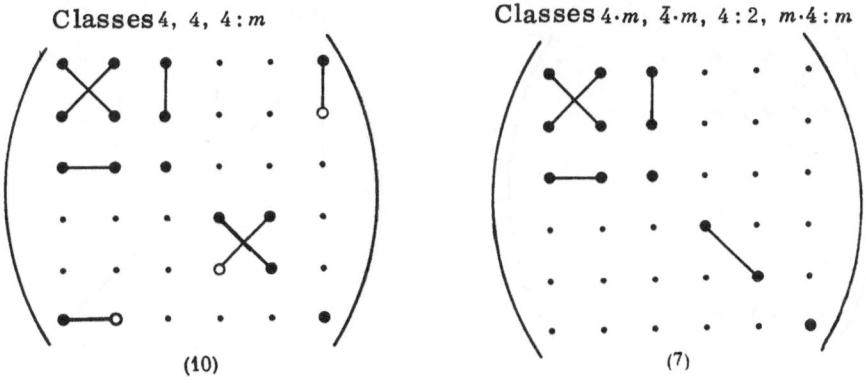

(10)

(7)

TABLE 62 (Continued)
Trigonal system

Classes 3, 6

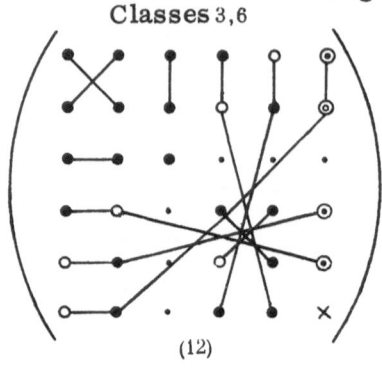

(12)

Classes 3·m, 3 : 2. 6·m

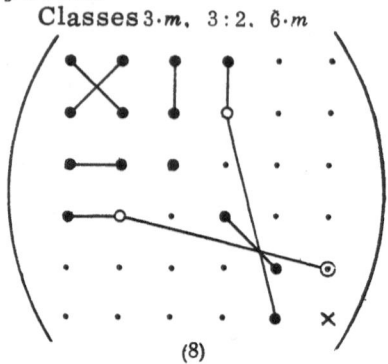

(8)

Hexagonal System and Some Textures With
∞ Symmetry Axes

Classes 6, 4, 6 : m, 3 : m
Textures ∞, ∞ : m

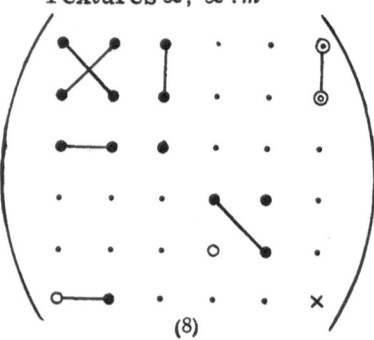

(8)

Classes m·3 : m, 6·m, 6 : 2, m·6 : m
Textures ∞·m, ∞ : 2, m·∞ : m

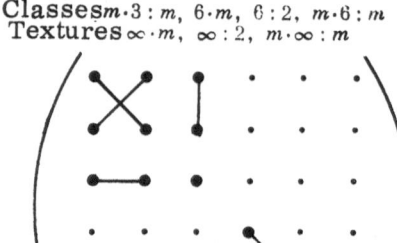

(16)

Cubic System

Classes 3/2; 6/2

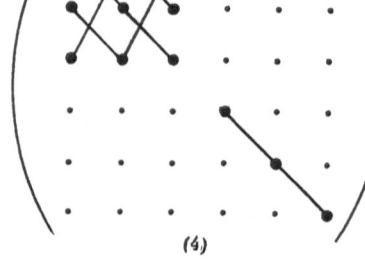

(4)

Classes 3/4, 3/4 и 6/4

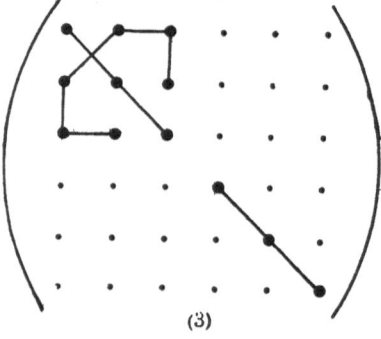

(3)

TABLE 62 (Continued)
Isotropic Media $\infty/m\cdot\infty : m$, $\infty/\infty : 2$

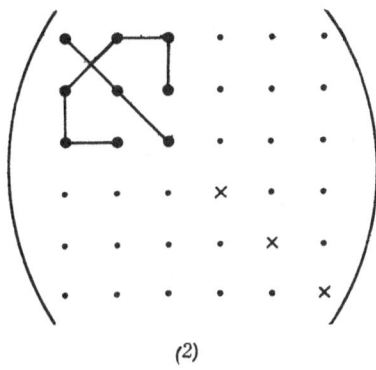

(2)

in Eqs. (IX.60a) and (IX.60b):

$$r_i = R_{ij}E_j^2,$$
$$r_{ij} = Q_{ij}P_j^2,$$
$$t_i = G_{ij}P_j^2,$$
$$t_i = H_{ij}P_j^2.$$

(IX.60c)

The tensor $[R_{ijmn}]$ and the other electrostrictive coefficients are similar to but not identical with the tensors of the elastic coefficients. The following permutations of the indices are permissible in the components of the electrostrictive coefficients:

$$R_{ijmn} = R_{ijnm} = R_{jimn} = R_{jimn},$$

but other permutations are not allowed:

$$R_{ijmn} \neq R_{mnij},$$

The matrices of the electrostrictive coefficients of all symmetry classes and of some textures are given in Table 62. This table lists matrices for the double-index forms of the coefficients R, Q, G, and H. When the components of the tensors $[r_{ij}]$ and $[t_{ij}]$ are reduced to the single-index form (§1 in the present chapter) and the products of the components of the vector P (or E) are used in the form given in the preceding paragraph, we find that the double-index forms of the electrostrictive coefficients can be expressed in terms of the coefficients with four indices:

$$\left. \begin{array}{l} G_{mn} = G_{ijkl} \\ H_{mn} = H_{ijkl} \end{array} \right\} \quad n = 1,\ 2,\ 3;$$

$$\left. \begin{array}{l} G_{mn} = 2G_{ijkl} \\ H_{mn} = 2H_{ijkl} \end{array} \right\} \quad n = 4,\ 5,\ 6;$$

$$\left. \begin{array}{l} R_{mn} = R_{ijkl} \\ Q_{mn} = Q_{ijkl} \end{array} \right\} \quad m \text{ and } n = 1,\ 2,\ 3;$$

$$\left. \begin{array}{l} R_{mn} = 2R_{ijkl} \\ Q_{mn} = 2Q_{ijkl} \end{array} \right\} \quad \text{either m or n equals 4, 5, or 6;}$$

$$\left. \begin{array}{l} R_{mn} = 4R_{ijkl} \\ Q_{mn} = 4Q_{ijkl} \end{array} \right\} \quad \text{m and n equals 4, 5, or 6.}$$

Application of Hooke's law to the strains r_{ij} and the stresses t_{ij} and the use of Eqs. (IX.60a) and (IX.60b) yields the following relationships between the electrostrictive coefficients:

$$\begin{array}{l} Q_{ijmn} = s^{P}_{ijkl} G_{klmn}, \\ G_{ijmn} = c^{P}_{ijkl} Q_{klmn}, \\ R_{ijmn} = s^{E}_{ijkl} H_{klmn}, \\ H_{ijmn} = c^{E}_{ijkl} G_{klmn}, \end{array} \qquad \text{(IX.61)}$$

in which the superscripts P and E of the elastic coefficients c and s indicate that these coefficients are measured under conditions of constant polarization or electric field, respectively. Moreover, allowance for the relationship between the polarization and the electric field

$$P_m = \alpha_{mn} E_n,$$

where α_{mn} is the polarizability tensor, gives the following relationships between the coefficients R and Q, as well as between G and H:

$$\begin{array}{l} R_{ijmn} = \alpha_{km} \alpha_{ln} Q_{ijkl}, \\ H_{ijmn} = \alpha_{km} \alpha_{ln} G_{ijkl}. \end{array} \qquad \text{(IX.62)}$$

Other relationships between the electrostrictive coefficients can also be obtained quite easily.

Electrostrictive stresses in crystals are second-order effects and, in general, they are small. Pure electrostriction can be observed only in nonpiezoelectric crystals. In piezoelectrics the strain of a crystal caused by the application of an electric field is due to the combined effect of piezoelectricity and electrostriction:

$$r_{jk} = d_{ijk}E_i + R_{iljk}E_i \cdot E_k,$$
$$r_{jk} = g_{ijk}P_i + Q_{iljk}P_i \cdot P_l. \qquad \text{(IX.63a)}$$

Rewriting these relationships in the form

$$r_{ij} = (d_{ijk} + Q_{iljk}E_l)\,E_i,$$
$$r_{ij} = (g_{iik} + Q_{iljk}P_l) \cdot P_l \qquad \text{(IX.63b)}$$

we can regard the second terms in parentheses as corrections to the piezoelectric coefficients, due to electrostriction caused by the application of an electric field. In this connection we must mention that, in general, the application of an electric field to a crystal can alter its symmetry and give rise not only to corrections but also new coefficients [see Eq. (IX.73)], corresponding to the new symmetry.

We shall show later that electrostrictive stresses in ferroelectrics can be large because of domain reorientation in electric fields.

B. Thermodynamic Relationships. We shall derive thermodynamic relationships describing electrostriction using a thermodynamic function known as the elastic enthalpy $H_1(t, D, S)$ (Table 49). In second-order effects, such as electrostriction, the role of thermal effects is slight and can be neglected. Therefore, we shall use only the stresses t_{ij} and the electric induction D_m as the independent variables.

Expanding the stress and electric field components as Maclaurin's series in terms of t and D and retaining terms up to the second order, we obtain the following expressions:

$$r_{ij} = \left(\frac{\partial r_{ij}}{\partial t_{kl}}\right)_D dt_{kl} + \left(\frac{\partial r_{ij}}{\partial D_n}\right)_t dD_n + \frac{1}{2!}\left[\frac{\partial^2 r_{ij}}{\partial t_{kl}\partial t_{qr}} dt_{kl}\,dt_{qr} + \right.$$

$$+ 2\frac{\partial^2 r_{ij}}{\partial t_{kl} \partial D_n} dt_{kl}\, dD_n + \frac{\partial^2 r_{ij}}{\partial D_n\, \partial D_0} dD_n\, dD_0\Big] + \ldots, \qquad \text{(IX.64)}$$

$$E_m = \left(\frac{\partial E_m}{\partial t_k}\right)_D dt_{kl} + \left(\frac{\partial E_m}{\partial D_n}\right)_t dD_n + \frac{1}{2!}\left[\frac{\partial^2 E_m}{\partial t_{kl} dt_{gr}} dt_{kl}\, dt_{gr} + \right.$$

$$\left. + 2\frac{\partial^2 E_m}{\partial t_{kl} \partial D_n} dt_{kl}\, dD_n + \frac{\partial^2 E_m}{\partial D_n\, \partial D_0} dD_n\, dD_0\right] + \ldots \qquad \text{(IX.65)}$$

Comparing Eqs. (IX.64) and (IX.29), we can see that the first two terms in parentheses are the elastic constants s^D and the piezoelectric coefficients g, respectively. The first term in the square brackets of Eq. (IX.64) describes changes in the elastic constants s under the action of mechanical stresses and the second term describes changes of the same elastic constants under the influence of the electric induction. These two effects will not be considered because they are not relevant to electrostriction.

The third term in the square brackets of Eq. (IX.64) describes electrostriction. We shall introduce the notation

$$(4\pi)^2 \frac{\partial^2 r_{ij}}{\partial D_n\, \partial D_0} = 2Q_{ijn0}, \qquad \text{(IX.66)}$$

where Q_{ijn0} is the electrostriction tensor of fourth rank.

The first two terms in parentheses of Eq. (IX.65) represent, respectively, the piezoelectric coefficients g and the reciprocals of the permittivity β^t [see Eq. (IX.30)]. The first term in the square brackets of Eq. (IX.65) describes changes in the piezoelectric coefficients g under the action of mechanical stresses and the third them describes changes in the reciprocals of the permittivities under the action of the electric induction. These two effects will not be considered here.

We can show that the second term in the square brackets of Eq. (IX.65) is equivalent to Eq. (IX.66). Using the thermodynamic potential H_1, we obtain

$$r_{ij} = -\frac{\partial H_1}{\partial t_{ij}}, \qquad E_m = 4\pi \frac{\partial H_1}{\partial D_m}. \qquad \text{(IX.67)}$$

Since the order of differentiation is unimportant, the partial derivatives are related by

$$4\pi \frac{\partial r_{ij}}{\partial D_n} = 4\pi \frac{\partial}{\partial D_n}\left(-\frac{\partial H_1}{\partial t_{ij}}\right) = -\ (4\pi)\frac{\partial^2 H_1}{\partial D_n\, \partial t_{ij}} = -\frac{\partial E_n}{\partial t_{ij}}. \quad \text{(IX.68)}$$

Taking the partial derivative of the last expression with respect to the electric induction and using Eq. (IX.66), we obtain

$$(4\pi)^2 \frac{\partial^2 r_{ij}}{\partial D_0\, \partial D_n} = -\ 4\pi \frac{\partial^3 H_1}{\partial D_n\, \partial D_0\, \partial t_{ij}} = -\ 4\pi \frac{\partial^2 En}{\partial t_{ij}\, \partial D_n} = 2Q_{ijno}. \quad \text{(IX.69)}$$

Thus, Eqs. (IX.64) and (IX.65), transformed so as to retain only the electrostrictive effect, become

$$r_{ij} = \frac{1}{(4\pi)^2} Q_{ijno} D_n D_0, \quad \text{(IX.70)}$$

where

$$E_m = -\frac{2}{4\pi} Q_{mnkl} D_n t_{kl}. \quad \text{(IX.71)}$$

It is interesting to interpret the equations for electrostriction in terms of the piezoelectric effect. Let us consider, for example, the case of electrostriction due to spontaneous polarization. Then, bearing in mind that under these conditions $D = 4\pi P$, we obtain from Eq. (IX.70):

$$r_{ij} = Q_{ijss} P_s^2. \quad \text{(IX.72)}$$

Taking the derivative of the last expression with respect to the polarization, we find that

$$\frac{\partial r_{ij}}{\partial P_s} = 2Q_{ijss} P_s,$$

and hence, using one of the equations of the system (IX.36), we obtain

$$g_{ijs} = 2Q_{ijss} P_s. \quad \text{(IX.73)}$$

Finally, Eq. (IX.70) can be written in the form

$$r_{ij} = \frac{1}{2} g_{ijs} P_s. \quad \text{(IX.74)}$$

The last expression allows us to treat electrostriction as a polarization-induced converse piezoelectric effect with a coefficient g proportional to the polarization [see Eq. (IX.73)]. This approach to the piezoelectric effect, which attributes it to the spontaneous polarization, has been used several times in the present chapter (see, for example, descriptions of properties of BaTiO$_3$ and triglycine sulfate in §3).

Considering Eq. (IX.71), we note, first of all, that it does not describe "converse" electrostriction. Such an effect does not exist at all, i.e., using only mechanical strains (or stresses) we cannot produce an electric polarization whose square is proportional to the applied strain or stress. This can be seen most easily in the case of centrosymmetric crystals which do not exhibit the linear (piezoelectric) effect but only the quadratic effect (electrostriction). Centrosymmetric mechanical stresses or strains do not alter the symmetry of centrosymmetric crystals and, therefore, we cannot expect any polarization (equivalent to the appearance of a unique polar direction).

Thus, Eq. (IX.71) describes a different phenomenon: it shows that the application of an electric field (or the establishment of polarization) imparts, even to centrosymmetric media, piezoelectric properties described by the coefficient g:

$$g_{nmk} = 2Q_{nmkl}P_l. \qquad\qquad (IX.75)$$

In this case, the direct piezoelectric effect (like the converse effect considered in the preceding sections) is the result of linearization of electrostriction: the piezoelectric coefficients are proportional to the value of the polarization. When a crystal is subjected to a mechanical stress a potential difference which appears between its electrodes is given by Eq. (IX.71):

$$E_m = - g_{mkl}t_{kl}, \qquad\qquad (IX.76)$$

which is identical with one of the equations in the system (IX.36).

Using other thermodynamic functions listed in Table 49, we can derive – in addition to Eq. (IX.70) – other equations for electrostriction which relate the quantities r and t with E and D. The complete system of equations for electrostriction can be written in the form

$$r_{ij} = \frac{1}{(4\pi)^2}\, Q_{ijmn} D_m D_n,$$

$$r_{ij} = R_{ijmn} E_m E_n,$$

$$t_{ij} = \frac{1}{(4\pi)^2}\, G_{ijmn} D_m D_n, \qquad \text{(IX.77)}$$

$$t_{ij} = H_{ijmn} E_m E_n,$$

where

$$(4\pi)^2 \frac{\partial^2 r_{ij}}{\partial D_m \partial D_n} = -\,4\pi\, \frac{\partial^2 E_m}{\partial t_{ij}\, \partial D_n} = 2 Q_{ijmn};$$

$$\frac{\partial^2 r_{ij}}{\partial E_m\, \partial E_n} = \frac{1}{4\pi}\, \frac{\partial^2 D_m}{\partial t_{ij}\, \partial E_n} = 2 R_{ijmn};$$

$$\frac{(4\pi)^2 \partial^2 t_{ij}}{\partial D_m \partial D_n} = 4\pi\, \frac{\partial^2 E_m}{\partial r_{ij}\, \partial D_n} = 2 G_{ijmn}; \qquad \text{(IX.78)}$$

$$\frac{\partial^2 t_{ij}}{\partial E_m\, \partial E_n} = \frac{1}{4\pi}\, \frac{\partial^2 D_m}{\partial r_{ij}\, \partial E_n} = 2 H_{ijmn}.$$

In the case of spontaneous electrostriction in a short-circuited crystal [see Eq. (IX.50)], the equations in the system (IX.77) transform to Eqs. (IX.60a) and (IX.60b):

$$r_{ij} = Q_{ijmn} P_m P_n,$$

$$r_{ij} = R_{ijmn} E_m E_n,$$

$$t_{ij} = G_{ijmn} P_m P_n, \qquad \text{(IX.79)}$$

$$t_{ij} = H_{ijmn} E_m E_n.$$

The relationships in Eq. (IX.78) change accordingly to:

$$\frac{\partial^2 r_{ij}}{\partial P_m\, \partial P_n} = -\,\frac{\partial^2 E_m}{\partial t_{ij}\, \partial P_n} = 2 Q_{ijmn},$$

$$\frac{\partial^2 r_{ij}}{\partial E_m\, \partial E_n} = \frac{\partial^2 P_m}{\partial t_{ij}\, \partial E_n} = 2 R_{ijmn},$$

$$\frac{\partial^2 t_{ij}}{\partial P_m\, \partial P_n} = \frac{\partial^2 E_m}{\partial r_{ij}\, \partial P_n} = 2 G_{ijmn}, \qquad \text{(IX.80)}$$

$$\frac{\partial^2 t_{ij}}{\partial E_m\, \partial E_n} = \frac{\partial^2 P_m}{\partial r_{ij}\, \partial E_n} = 2 H_{ijmn}.$$

C. Physical Meaning and Methods for Determination of Electrostrictive Coefficients. The physical meaning of the coefficients Q, R, G, and H follows directly from the system of equations (IX.79). Thus, for example, the coefficient R represents the strain in a free crystal subjected to an electric field, and so on. These coefficients are related to each other by dielectric and elastic parameters of a crystal and they represent the operation of electrostrictive transducers under various conditions.

The electrostrictive coefficients can be measured directly. Thus, the coefficient Q is found by measuring the strain generated by the applied field, the coefficient H is calculated by measuring the mechanical stress and comparing it with the square of the applied field E, and so on. The experimental conditions used to determine the coefficients Q, R, G, and H from the system of equations (IX.77) are given in Fig. 239. Comparison of the electrostrictive and converse piezoelectric effects and the examination of this figure show that the conditions for the determination of the coefficient Q are equivalent to the conditions necessary to determine the piezoelectric coefficient g. The coefficient R corresponds to the piezoelectric strain coefficient d, G corresponds to h, and H corresponds to e.

Apart from direct methods for determination of the electrostrictive coefficients, there are also some indirect methods. These methods follow from the expressions in Eq. (IX.78), which define the electrostrictive coefficients. It is important to note that there are two indirect methods for determination of each of the electrostrictive coefficients. We can illustrate this by considering the coefficients Q and R, since they are used most widely. The coefficient Q can be determined by: a) measuring the dependence of the piezoelectric coefficient g on the applied field; b) measuring the dependence of the reciprocal of the permittivity on the mechanical stress. The coefficient R can be determined by measuring the dependence of the piezoelectric strain coefficient d on the electric field or the dependence of the permittivity on the mechanical stress.

D. Electrostriction of Linear Dielectric Crystals. Electrostrictive strains in crystals of linear dielectrics are extremely small. This is why, until recently, there have been no published values of the electrostrictive coefficients of crystals.

A crystal of quartz does not exhibit the piezoelectric effect when an electric field is applied to it along the Z axis [Eq. (IX.43)].

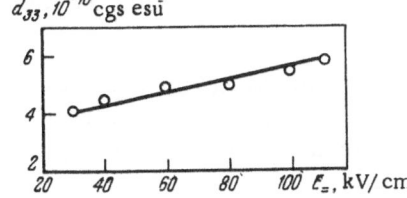

Fig. 261. Dependence of the induced piezoelectric strain coefficient d_{33} of Z-cut quartz
on the static field intensity [4].

Thus, an electric field applied along the Z axis of quartz can only
produce an electrostrictive strain. The electrostrictive coefficient
R_{33} of quartz has been determined by obtaining the dependence of
the piezoelectric effect induced by a static field (the piezoelectric strain coefficient d_{33}) on the value of this field applied along
the Z axis (Fig. 261). In this determination the value of the piezoelectric strain coefficient is measured in various static fields by
the application of a weak alternating field. Using the second equation in the system (IX.78), we obtain

$$d_{33} = 2R_{33}E_3. \qquad (IX.81)$$

The value of R_{33}, found from the slope of the straight line in Fig.
261, is 0.3×10^{-12} cgs esu.

It is evident from Fig. 261 that the value of d_{33} in a field $E_3 =$
100 kV/cm is 2×10^{-10} cgs esu (after subtracting the "constant
component," which will be discussed later). This value is approximately two orders of magnitude lower than the value of d_{11} of quartz
($\sim 6.8 \times 10^{-8}$ cgs esu). It follows that a field of the order of 10^7
V/cm would have to be applied to a Z-cut quartz crystal in order
to obtain a piezoelectric strain coefficient numerically equal to d_{11}.
The straight line in Fig. 261 does not pass through the origin of
coordinates because of some inaccuracy in the orientation of the
Z-cut plate. We can easily see that such an inaccuracy, i.e., a deviation of the plane of the plate from an orientation strictly per-

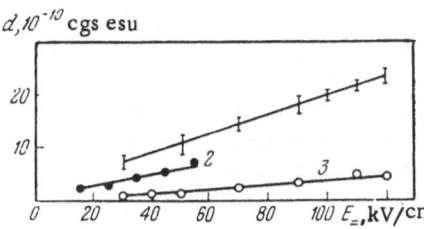

Fig. 262. Dependence of the induced piezoelectric strain coefficients of NaCl on the
static field intensity: 1) d_{33}; 2) d_{31}; 3) d'_{33}
(displacement along the [111] direction) [4].

pendicular to the Z axis, gives rise to linear effects which are not
due to the applied static field.

The electrostrictive coefficients of rocksalt (Fig. 262) have
been measured in a similar manner. The values of the electrostrictive coefficients found from the dependences d(E) for rocksalt
are:

$$R_{11} = R_{22} = R_{33} = 2\ 7 \cdot 10^{-12},$$
$$R_{12} = R_{13} = R_{23} = -1.35 \cdot 10^{-12},$$
$$R_{44} = 0.9 \cdot 10^{-12} \text{ cgs esu.}$$

Comparison of the data for quartz and rocksalt shows that
the electrostriction of rocksalt is much stronger than that of quartz.
For example, one of the piezoelectric strain coefficients of rocksalt is 2×10^{-9} cgs esu in a field of 100 kV/cm.

E. Electrostriction of Ferroelectric Crys-
tals. The values of the electrostrictive coefficients of BaTiO$_3$
crystals, estimated using various methods (e.g., from the sudden
change in the unit cell parameters and the spontaneous polarization) are in good agreement. Their values are (Chap. V):

$$Q_{11} = 1.7 \cdot 10^{-12},$$
$$Q_{12} = -0.53 \cdot 10^{-12},$$
$$Q_{44} = 0.70 \cdot 10^{-12} \text{ cgs esu.}$$

It follows from the conditions under which the coefficients Q
are measured (these conditions are equivalent to those employed
in the determination of the piezoelectric coefficients g) and from
factors analogous to those which govern the behavior of the coefficients g in the phase transition region that the electrostrictive coefficients Q and G should exhibit no anomalies near the

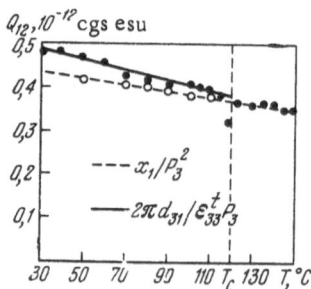

Fig. 263. Temperature dependence of the electrostrictive coefficient Q$_{12}$ of BaTiO$_3$ [17].

Point Group Symbols*

Crystallographic System	Notation System		Schoenflies
	Shubnikov	International	
Triclinic	1	1	C_1
	$\bar{2}$	$\bar{1}$	$C_i = S_2$
Monoclinic	2	2	C_2
	m	m	$R_{1h} = C_5$
	$2{:}m$	$2/m$	C_{2h}
Orthorhombic	$2{:}2$	222	$D_2 = V$
	$2{\cdot}m$	$mm2$	C_{2v}
	$m{\cdot}2{:}m$	mmm	$D_{2h} = V_h$
Rhombohedral	4	4	C_4
	$\bar{4}$	$\bar{4}$	S_4
	$4{:}2$	422	D_4
	$4{\cdot}m$	$4mm$	C_{4v}
	$4{:}m$	$4/m$	C_{4h}
	$\bar{4}{\cdot}m$	$\bar{4}2m$	$D_{2d} = V_d$
	$m{\cdot}4{:}m$	$4/mmm$	D_{4h}
Hexagonal	3	3	C_3
	$\bar{6}$	$\bar{3}$	$C_{3i} = S_6$
	$3{:}2$	$3/2$	D_3
	$3{\cdot}m$	$3m$	C_{3v}
	$\bar{6}{\cdot}m$	$\bar{3}/m$	D_{3d}
	6	6	C_6
	$3{:}m$	$\bar{6}$	C_{3h}
	$6{:}2$	622	D_6
	$6{\cdot}m$	$6mm$	C_{6v}
	$6{:}m$	$6/m$	C_{6h}
	$m{\cdot}3{:}m$	$\bar{6}2$	D_{3h}
	$m{\cdot}6{:}m$	$6/mmm$	D_{6h}
Cubic	$3/2$	23	I
	$3/4$	432	O
	$3/\bar{4}$	$\bar{4}3m$	I_d
	$\bar{6}/2$	$m3$	I_h
	$\bar{6}/4$	$m3m$	O_h

*The Shubnikov notation is used throughout this book; the table above gives the International and Schoenflies symbols for the convenience of readers. Keep this page folded out if you require quick conversion from the Shubnikov to the other systems. The different notation systems are also given in Table 3 on pp. 12-13 in Vol. 1.

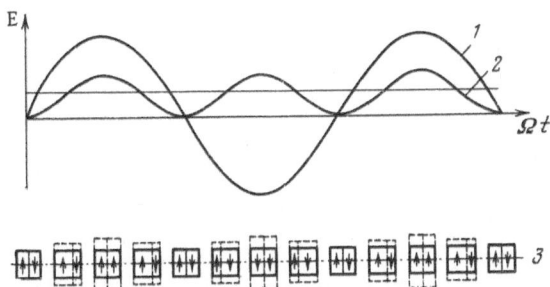

Fig. 264. Schematic representation of strain in a ferroelectric subjected to an alternating electric field: 1)
time dependence of the field; 2, 3) electrostrictive strain.

phase transition point. This is supported by the experimental data (Fig. 263). However, anomalies of the electrostrictive coefficients R and H should be observed near the Curie point.

No direct experimental determination has yet been carried
out of the electrostrictive coefficients of KH_2PO_4. A comparison of the spontaneous polarization and spontaneous strain in the
region of the Curie point can be used to estimate the value of the
coefficient Q_{33} (Chap. V):

$$Q_{33} = 4.3 \cdot 10^{-12} \text{ cgs esu.}$$

The cited values of the electrostrictive coefficients of $BaTiO_3$
and KH_2PO_4 apply to single-domain crystals. The presence of domains and their reorientation (in alternating fields used to measure electrostriction) increase strongly the electrostrictive coefficients. The domain contribution to electrostriction is basically due to the converse piezoelectric effect because of double switching of domains in each period of the field (Fig. 264). Under real
conditions the electrostrictive vibrations (displacements) are ex-
cited by an alternating field $E_\sim$. This case is represented by the
electrostrictive coefficient R and involves domain switching. The
simultaneous application of a static (polarizing) and an alternating field "suppresses" the domain mechanism and reduces electrostriction (this will be discussed later).

A comparison of the spontaneous strain and spontaneous polarization gives the coefficient $Q_{22} = -15.0 \times 10^{-12}$ cgs esu for triglycine sulfate crystals.

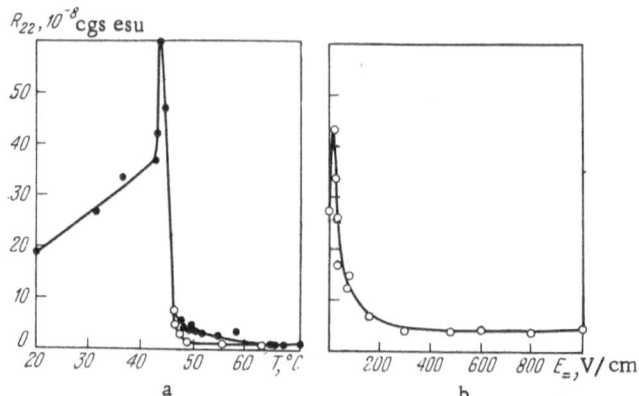

Fig. 265. Electrostrictive coefficient R_{22} of triglycine sulfate as a function of temperature (a) and of polarizing field applied at 26.7°C (b); measurements in an alternating field of 1 kc frequency [14].

The results of an experimental investigation of the electrostrictive coefficient R_{22} of triglycine sulfate are presented in Fig. 265. The anomaly of R_{22} in the phase transition region (Fig. 265a) is associated with the anomaly of the permittivity ε_2. The value of R_{22} of triglycine sulfate decreases rapidly at 47.3°C, i.e., at a temperature which is slightly lower than the Curie point (~49°C). The room-temperature value of R_{22} is 18×10^{-8} cgs esu, the value immediately after the rapid drop (at 47.3°C) is 5×10^{-8} cgs esu, and at 60-70°C this coefficient is 1×10^{-8} cgs esu. Using one of the relationships between the electrostrictive coefficients,

$$R_{22} = \frac{e_z^2}{(4\pi)^2} Q_{22}, \qquad (\text{IX.82})$$

and the value $\varepsilon = 200$ at 70°C (Fig. 119), we find that the coefficient R_{22} at this temperature (1×10^{-8} cgs esu) yields $Q_{22} = -36 \times 10^{-12}$ cgs esu, which is in satisfactory agreement with the value cited in the preceding paragraph, when we bear in mind that other measurements on triglycine sulfate yield $Q_{22} = -27 \times 10^{-12}$ cgs esu.

The dependence of the electrostrictive coefficient R_{22} on the static field intensity $E_=$ (Fig. 265b) represents the process of gradual "exclusion" of domains from electrostriction with increasing value of $E_=$. The fact that the maximum value of R_{22} does not correspond to $E_= = 0$ indicates that the highest mobility of domains is observed in some finite bias field (in general, this field is close to the coercive value, as discussed in the section on nonlinear phenomena in Chap. VII).

The electrostrictive coefficients Q_{ij} of Rochelle salt, found from direct measurements of thermal expansion, are (Chap. V):

$$Q_{11} = -86.5 \cdot 10^{-12},$$
$$Q_{12} = 17.3 \cdot 10^{-12},$$
$$Q_{13} = -24.3 \cdot 10^{-12} \text{ cgs esu.}$$

The values obtained by other methods are in good agreement with each other.

The temperature dependence of the electrostrictive coefficient R_{11} of Rochelle salt is shown in Fig. 266a. The rapid rise of this coefficient near the Curie point, similar to that observed for triglycine sulfate, is associated with the anomaly of ε_1. It is evident from this figure that up to 21°C the electrostrictive coefficient increases with increasing field. However, in the temperature range 21-23°C the electrostrictive coefficient, measured using an alternating field of 90 V/cm intensity, is higher than in a field of 110 V/cm. It follows that (at this temperature and fre-

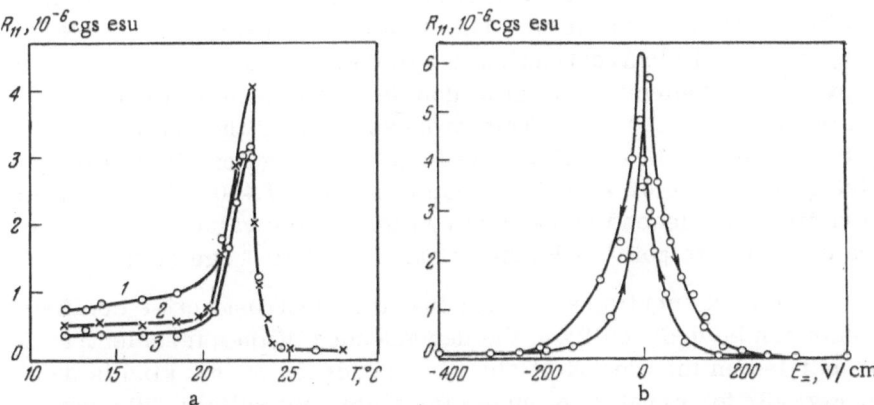

Fig. 266. Electrostrictive coefficient R_{11} of Rochelle salt. a) Temperature dependence of R_{11} in alternating fields $E_\sim$ of various intensities (V/cm): 1) 110; 2) 90; 3) 70; b) dependence of R_{11} on the polarizing field intensity (measurements in $E_\sim = 140$ 140 V/cm at 12°C) [12].

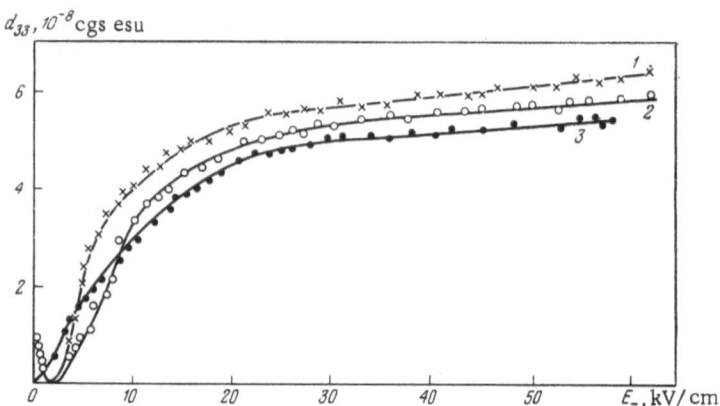

Fig. 267. Dependence of the piezoelectric strain coefficient d_{33} of Z-cut guanidine aluminum sulfate on the intensity of a polarizing field for unipolar (1 and 2) and nonunipolar samples (3) [2].

quency) a field of 90 V/cm is capable of switching the majority of domains so that a further increase of the field intensity produces a relatively smaller increase in the number of switched domains. Above 23°C the electrostrictive coefficient is the same for all three alternating field intensities. The value of the coefficient R_{11} at 28°C is 0.1×10^{-6} cgs esu.

Figure 266b shows the dependence of the electrostrictive coefficient R_{11} of a Rochelle salt crystal (ferroelectric cut) on the applied static electric field. This figure illustrates clearly the suppression of electrostriction due to domain reorientation (the converse piezoelectric effect at double the frequency) when the external field is increased. This process reduces the coefficient R_{11} quite strongly: the coefficient is 6.15×10^{-6} cgs esu in a zero static field and 0.07×10^{-6} cgs esu in a field of 400 V/cm of either polarity. The second value is very close to the value of R_{11} above the upper Curie point of Rochelle salt (0.1×10^{-6} cgs esu).

We have mentioned earlier that the electrostrictive coefficients can be deduced from the dependences of the piezoelectric properties on the electric field. A dependence of this kind is shown in Fig. 267 for crystals of guanidine aluminum sulfate. We can easily see that curve 3 gives, after differentiation, a dependence with a maximum, similar to that shown in Fig. 265b for triglycine sulfate. The value of the electrostrictive coefficient R_{33} [Eq.

(IX.81)] of guanidine aluminum sulfate, calculated in the range of static fields corresponding to saturation, is 20×10^{-12} cgs esu. This value is more than three orders of magnitude lower than the value of R_{11} characterizing electrostriction of Rochelle salt in fields close to saturation. Such a large difference between the coefficients R of these two crystals is evidently due to different values of the temperature interval between the measurement temperature and the Curie temperature and due to the higher polarizing field in the case of guanidine aluminum sulfate.

The value of R of guanidine aluminum sulfate can be used to calculate the electrostrictive coefficient Q using Eq. (IX.82); such a calculation gives $Q_{33} = 13.3 \times 10^{-12}$ cgs esu, which is similar to the values of Q for other ferroelectric crystals.

Figure 267 is also interesting because it shows the process of polarization of a guanidine aluminum sulfate crystal (represented by an increase of the piezoelectric strain coefficient d_{33}); the data given in the figure can be used to estimate the coercive field, saturation field, etc. Moreover, extrapolation of the linear part of the dependence $d(E_=)$ to $d_{33} = 0$ can be used to estimate the field which must be applied to a guanidine aluminum sulfate crystal in order to produce a polarization equal to the spontaneous value. An estimate of this field gives 2.7×10^5 V/cm. This is much lower than the fields required to produce a piezoelectric effect in quartz of a value similar to the natural effect.

References

1. I. S. Zheludev, Kristallografiya, 2:89 (1957).
2. I. S. Zheludev and V. S. Lelekov, Kristallografiya, 7:463 (1962).
3. I. S. Zheludev and N. A. Romanyuk, Kristallografiya, 4:710 (1959).
4. I. S. Zheludev and A. A. Fotchenkov, Kristallografiya, 3:308 (1958).
5. V. A. Koptsik, Kristallografiya, 4:219 (1959).
6. V. A. Koptsik and I. B. Kobyakov, Kristallografiya, 4:223 (1959).
7. W. G. Cady, Piezoelectricity, McGraw-Hill, New York (1946); rev. ed. Dover, New York (1964).
8. W. P. Mason, Piezoelectric Crystals and Their Applications to Ultrasonics, Van Nostrand, New York (1950).
9. J. F. Nye, Physical Properties of Crystals, Clarendon Press, Oxford (1957).
10. I. M. Sil'vestrova, Dissertation [in Russian], Institut Kristallografii AN SSSR, Moscow (1963).
11. A. A. Fotchenkov, Kristallografiya, 5:415 (1960).

12. A. A. Fotchenkov, I. S. Zheludev, and M. P. Zaitseva, Kristallografiya, 6:576 (1961).

13. A. A. Fotchenkov and M. P. Zaitseva, Kristallografiya, 7:934 (1962).

14. A. A. Fotchenkov, M. P. Zaitseva, and L. I. Zherebtsova, Kristallografiya, 8:724 (1963).

15. A. A. Chumakov, I. M. Sil'vestrova, and K. S. Aleksandrov, Kristallografiya, 2:707 (1957).

16. D. Berlincourt and H. Jaffe, Phys. Rev., 111:143 (1958).

17. E. J. Huibregtse, W. H. Bessey, and M. E. Drougard, J. Appl. Phys., 30:899 (1959).

18. E. J. Huibregtse, M. E. Drougard, and D. R. Young, Phys. Rev., 98:1562 (1955).

19. H. J. Martin, Die Ferroelektrika, Geest and Portig, Leipzig (1964).

20. V. Petržilka, J. B. Slavik, I. Šolc, O. Taraba, J. Tichý, and J. Zelenka, Piezo-elektřina a Její Technicke Použití, Jednoty českych matematiku a fysiku, Prague (1960).

21. J. Valasek, Phys. Rev., 19:478 (1922).

22. F. Jona and G. Shirane, Ferroelectric Crystals, Pergamon Press, Oxford (1962).

23. W. Känzig, "Ferroelectrics and antiferroelectrics," Solid State Phys., 4:1 (1957).

24. V. P. Konstantinova, I. M. Sil'vestrova, and K. S. Aleksandrov, Kristallografiya, 4:69 (1969).

25. V. P. Konstantinova, I. M. Sil'vestrova, and K. S. Aleksandrov, Physics of Dielectrics (Proc. Second All-Union Conf., Moscow 1958) [in Russian], Izd. AN SSSR, Moscow (1960), p. 31.

26. A. A. Kharkevich, Theory of Transducers [in Russian], Gosénergoizdat, Moscow-Leningrad (1948).

27. A. V. Shubnikov, Quartz and Its Applications [in Russian], Izd. AN SSSR, Moscow (1940).

28. V. A. Yurin, I. M. Sil'vestrova, and I. S. Zheludev, Kristallografiya, 7:394 (1962).

Index*

---

*pp. 1-336, inclusive, are in Volume 1; pp. 337-620 are in Volume 2.

xxi